Mining Techniques - Past, Present and Future

Edited by Abhay Soni

Published in London, United Kingdom

IntechOpen

Supporting open minds since 2005

Mining Techniques - Past, Present and Future
http://dx.doi.org/10.5772/intechopen.79241
Edited by Abhay Soni

Contributors
Xiaohan Yang, Vladimír Sedlák, Rami Alrawashdeh, Awwad Altiti, Hani Alnawafleh, Ashok Kumar, Dheeraj Kumar, Arun Kumar Singh, Sahendra Ram, Rakesh Kumar, Mudassar Raja, Amit Kumar Singh, Luiz Fernando Spinelli Pinto, Lizete Stumpf, Pablo Miguel, Leonir Aldrighi Dutra Junior, Jeferson Diego Leidemer, Lucas Da Silva Barbosa, Mauricio Silva E Oliveira, Yıldırım İsmail İsmail Tosun, Paweł Kamiński, Piotr Czaja, Artur Dyczko, Abhay Soni, R.D. Dwivedi

First published in London, United Kingdom, 2021 by IntechOpen
IntechOpen is the global imprint of INTECHOPEN LIMITED, registered in England and Wales, registration number: 11086078, 5 Princes Gate Court, London, SW7 2QJ, United Kingdom
Printed in Croatia

British Library Cataloguing-in-Publication Data
A catalogue record for this book is available from the British Library

Additional hard and PDF copies can be obtained from orders@intechopen.com

Mining Techniques - Past, Present and Future
Edited by Abhay Soni
p. cm.
Print ISBN 978-1-83962-368-4
Online ISBN 978-1-83962-369-1
eBook (PDF) ISBN 978-1-83962-370-7

For EU product safety concerns:
IN TECH d.o.o., Prolaz Marije Krucifikse Kozulić 3, 51000 Rijeka, Croatia, info@intechopen.com or visit our website at intechopen.com.

Meet the editor

Dr. A.K. Soni, Ph.D., graduated with a degree in Mining Engineering from Ravishankar University, Raipur, Chhattisgarh, in 1983. He completed his post-graduate studies at the Birla Institute of Technology and Science (BITS), Rajasthan, India, and obtained a Ph.D. in Environmental Science and Engineering from the Centre of Mining Environment, Indian School of Mines (ISM), Dhanbad, India, in 1998. Dr. Soni is currently working as Chief Scientist at CSIR-Central Institute of Mining and Fuel Research (CSIR-CIMFR) at Nagpur Research Centre and engaged in research work on "mine environment and allied areas." His area of research interest is "geo-hydrological problems related to mines." He has more than 33 years of experience working in the Indian mining industry. As part of his research work, he has visited the United States and the United Kingdom and traveled widely across India. As a research scientist and technical administrator, he has more than 115 technical publications on mining and environmental topics to his credit. Dr. Soni has authored one book, Mining in the Himalayas: An Integrated Strategy. He has also written technical papers in the Hindi language. Dr. Soni has handled more than 100 R&D projects in the capacity of project coordinator and principal investigator. He is actively associated with professional societies in India, including the Mining Engineers Association of India (MEAI), Institution of Engineers (India), Indian Society for Rock Mechanics and Tunneling Technology (ISRMTT), and International Mine Water Association (IMWA). Dr. Soni has received many honors and awards for his contributions. He is presently a member of the international advisory board for the Journal of Mine Water and Environment. He is also a member and chairman of important committees, and a subject area expert, advisor, and evaluator responsible for several noted professional assignments at the national level. He has been invited by academic institutes and Indian universities to deliver lectures and conduct examinations for post-graduate students. Dr. Soni was associated with the Bureau of Indian Standards (BIS) in the capacity of member and has experience organizing several technical events.

Contents

Preface

Mining is an ancient activity that began at the inception of civilization. On re-examining the mining activities of the present day, it is apparent that several transformations in mining techniques have occurred. This book examines the past, present, and future missions of mining. The growing needs of the mining industry have led to larger investments in the field of "mining science," which encompasses engineering, mechanics, chemistry, physics, and management. In the next few decades, "conventional methods" will be replaced by advanced and modern techniques of fast and safe excavations involving technology, machinery, and equipment.

Written by authors from Poland, India, Australia, Brazil, Turkey, Jordan, and Slovakia, this volume includes nine chapters over two sections. Section 1 examines current mining methods and techniques such as open-pit mining and hill mining. Other chapters in this section examine environmental degradation and reclamation due to surface coal mining as well as mining in hills and tunneling.

Section 2 includes chapters on future mining techniques such as shaft excavation, which is examined using a case study of the shaft-sinking experiences of Polish engineers at the Leon IV shaft of the Rydułtowy Mine, Poland.

Further, 'mineral resources' are hidden treasures that can give a boost to commercial activities, the industry as well as entrepreneurs. Activities related to these resources can generate employment and undoubtedly make a significant contribution to GDP growth for mineral-rich states and countries. As far as my optimistic understanding of S&T, R&D and mining system and subject as a whole goes, I believe that this book will be useful for all professionals of the mining industry.

This book is an important resource on mining, minerals, the environment, and engineering. Its applied research concepts and comprehensive content will be beneficial for researchers, students, policy formulators, and decision-makers in different areas of the mining industry. Mine operators and civil or excavation engineering professionals will also find the information presented herein useful and practical.

Compiling and editing a book on the past, present, and future of various mining techniques is a stupendous task, as the mining industry worldwide is extensive and varied. Therefore this book does not cover every aspect of mining; that is a challenge for a future book.

As an editor, I would like to thank those who assisted me with this book project, particularly my organization (CSIR-Central Institute of Mining and Fuel Research) and all my colleagues, of which there are too many to name. I am indebted to them for all their help.

Dr. A.K. Soni
Chief Scientist,
Mining Technology Group,
Nagpur Research Centre,
CSIR-Central Institute of Mining and Fuel Research (CSIR-CIMFR),
Nagpur, India

Section 1

Mining Techniques - Past and Present

Chapter 1

Open Pit Mining

Awwad H. Altiti, Rami O. Alrawashdeh and Hani M. Alnawafleh

Abstract

Open pit mining method is one of the surface mining methods that has a traditional cone-shaped excavation and is usually employed to exploit a near-surface, nonselective and low-grade zones deposits. It often results in high productivity and requires large capital investments, low operating costs, and good safety conditions. The main topics that will be discussed in this chapter will include an introduction into the general features of open pit mining, ore body characteristics and configurations, stripping ratios and stripping overburden methods, mine elements and parameters, open pit operation cycle, pit slope angle, stability of mine slopes, types of highwall failures, mine closure and reclamation, and different variants of surface mining methods including opencast mining, mountainous mining, and artisan mining.

Keywords: open pit mine, slope stability, mine reclamation, stripping ratio, production cycle

1. Introduction

Open pit mining is defined as the method of extracting any near surface ore deposit using one or more horizontal benches to extract the ore while dumping overburden and tailings (waste) at a specified disposal site outside the final pit boundary. Open pit mining is used for the extraction of both metallic and nonmetallic ores. Open pit mining is considered different from quarrying in the sense that it selectively extracts ore rather than an aggregate or a dimensional stone product.

Open pit mining is applied to disseminated ore bodies or steeply dipping veins or seams where the mining advance is toward increasing depths. Backfilling usually occurs until the pit is completed; even then, the high cost of filling these pits with all of the waste removed at the end of the mine life would seriously risk the project's economics. Few large open pits in the world could support such a costly obstacle. Open pit method is usually nonselective, and it includes all high and low-grade zones; whereas mining rate is nearly over 20,000 tons mined per day and often necessitates a large capital investment but generally results in high productivity, low operating cost, and good safety conditions [1]. The main purpose of this chapter is to discuss the general features of open pit mining, ore body characteristics and configurations, stripping ratios and stripping overburden methods, mine elements and parameters, open pit operation cycle, pit slope angle, stability of mine slopes, types of highwall failures, mine closure, and reclamation. The chapter will also discuss different variants of surface mining methods including opencast mining, mountainous mining, and artisan mining.

1.1 Features, technical and economic indicators of open pit development

Compared to underground mining methods, the open pit mining method requires removing significant amount of overburden from the pit and moving it outside the mine. The cost of extraction of the ore from open pit constitutes the bulk of the total cost of mining operations, because the access to the ore body is so fast and requires less time compared to underground mining, i.e., extracting the ore below overburden can only begin with some lag time from the start of removing overburden. Also, open pit has virtually an unlimited ability to create and use high-performance large-sized mining and transportation equipment that can provide the highest technical and economic parameters. Open pit mining has higher productivity (3–5 times of underground methods), lower production costs, more safe and hygienic working conditions, more complete recovery of a mineral, and lower per unit production cost.

Open pit mining is characterized not only by its high share of total minerals production, but it is also considered as one of the surface mining methods that contributes to the construction of powerful performance quarries (100–150 million tons of rock a year reaching to a depth of 500 m). Capital cost of such huge open pits/quarries is very high, and the total cost for excavation of rock in the long term reaches hundreds of millions of dollars or more. Therefore, decisions on the construction of new or existing quarries should be economically justified. **Table 1** shows the advantages and disadvantages of open pit mining method [2, 3].

1.2 Ore body characteristics and configurations

Open pit mining is widely used with metallic ore bodies (aluminum, bauxite, copper, iron), and nearly all nonmetallic (coal, uranium, phosphate, etc.). It is a traditional cone-shaped excavation (although it can be of any shape, depending on the size and shape of the ore body) that is used when the ore body is typically pipe-shaped, vein-type, and steeply dipping stratified or irregular [4]. The major open pit and ore body configurations are classified into the following:

- Flat lying seam or bed, flat terrain (e.g., platinum reefs, coal), as shown in **Figure 1**.

- Massive deposit, flat terrain (e.g., iron-ore or sulfide deposits), as shown in **Figure 2**.

- Dipping seam or bed, flat terrain (e.g., anthracite), as shown in **Figure 3**.

- Massive deposit, high relief (e.g., copper sulfide), as shown in **Figure 4**.

- Thick-bedded deposits, little overburden (e.g., iron ore, coal) as shown in **Figure 5**.

1.3 Stripping ratio

The parameter known as the *stripping ratio* represents the amount of uneconomic material that must be removed to uncover one unit of ore, i.e., the ratio of the number of tons of waste material removed to the number of tons of ore removed. Also, the ratio of the total volume of waste to the total volume of ore is defined as the *overall stripping ratio.* A lower stripping ratio means that less waste has to be

Advantages	Disadvantages
High productivity, i.e., highly mechanized and labor conserving (around 100–400 tons per employee-shift including both ore and waste)	Limited by depth ~500 m; technological limit imposed by equipment; and deposit beyond pit limits must be mined underground or left in place
High production rate (essentially unlimited, although small surface mines also possible)	Limited by stripping ratio
Lowest cost along with open cast mining	High capital investment associated with large equipment
Early production, development can be programmed to permit early start-up	Surface damaged may require reclamation; a bond has to be added to the production cost
Low labor requirement; can be unskilled except key operators (e.g., drill, shovel)	Requires large deposit to realize lowest cost, unless very high grade
Relatively flexible; can vary output if demand changes	Weather can be detrimental; it can impede operations.
Suitable for large equipment; permit high productivity	Slope stability must be maintained; proper design and maintenance of benches plus good drainage are essential
Fairly low rock-breakage cost (drilling and blasting); superior to underground mining where bench faces are less easily maintained	Requires provision of large waste disposal/dump area
Simple development and access; minimal openings required although advanced stripping may be considerable	
Little if any bank support required; proper design and maintenance of benches can provide stability	
Good recovery; good health and safety; no underground hazards	

Table 1.
Advantages and disadvantages of open pit mining method [2, 3].

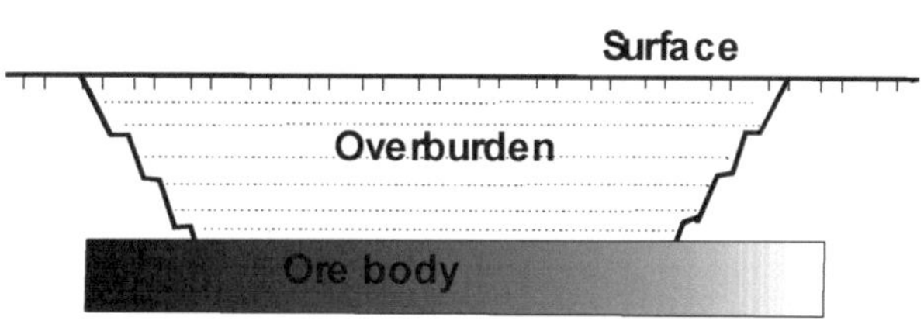

Figure 1.
Flat lying seam or bed, flat terrain.

removed to expose the ore for mining which generally results in a lower operating cost [5]. The major types of stripping ratios are overall, instantaneous, and break-even.

In order to specify the maximum allowable stripping ratio (SR_{max}) of a surface mine, break even ratio can help to establishes the pit limits. SR_{max} defined as the ratio of overburden to ore at the ultimate boundary of the pit, where the profit margin is zero. It can be calculated as:

$$SR_{max} = (\text{Value of ore} - \text{Production cost})/\text{Stripping cost} \tag{1}$$

or,

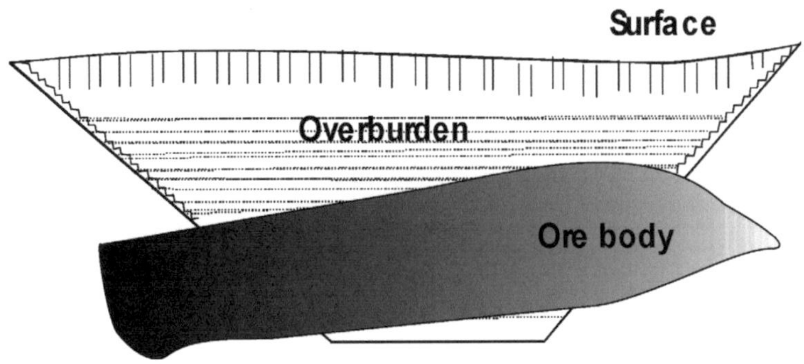

Figure 2.
Massive deposit, flat terrain.

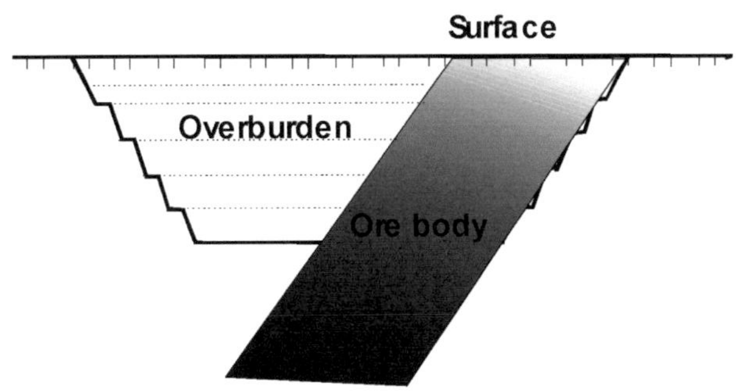

Figure 3.
Dipping seam or bed, flat terrain.

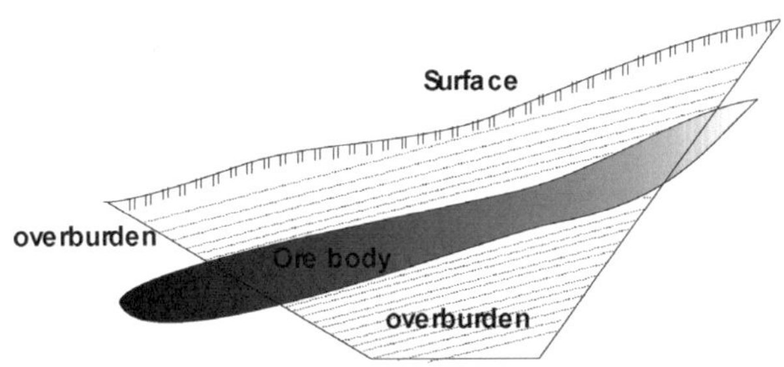

Figure 4.
Massive deposit, high relief.

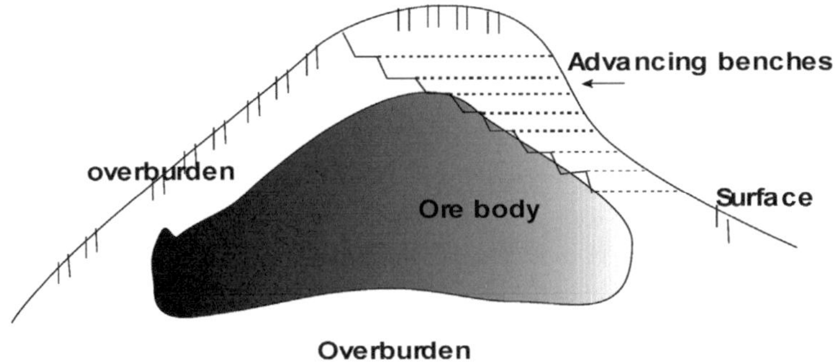

Figure 5.
Thick-bedded deposits.

$$SR_{max} = \text{Stripping allowance } (\$/\text{ton})/\text{Stripping cost } \left(m^3/\text{ton}\right) \tag{2}$$

The maximum allowable stripping ratio enables us to locate the ultimate pit boundary or limit based on prevailing economic, physical, and geometric conditions in the pit. A copper pit designed in this manner with varying ore grades and critical $SR_{max} \sim 2.5\ m^3/ton$ ($3.0\ yd^3/ton$) is shown in **Figure 6**. Ore occurring in the ore body beyond this maximum stripping ratio will have to be left or mined underground.

Example 1:

A seam of coal has a density of 1.36 t/m^3 and is 2.5 m thick. It is covered by 27 m of shale which has a density of 1.7 t/m^3. Calculate the stripping ratios?

Solution:

- *Volumetric stripping ratio* = 27/2.5 =10.8 m^3 of overburden per m^3 ore
- *Weight stripping ratio* = (27 × 1.7)/(2.5 × 1.36) = 13.5 tons of overburden per ton ore
- *Stripping ratio* = 27/(2.5 × 1.36) = 7.9 m^3 of overburden per ton ore

Example 2:

The head assay of a copper ore is 0.8% Cu. The expected overall copper recovery from the ore is 88%. Calculate the maximum stripping ratio if the total cost of production (excluding overburden removal) is \$5.90 per ton of ore and overburden removal costs are \$0.3 per ton of waste. Assume copper values of \$1.00, \$1.25, and \$1.50 per kg of refined metal at the smelter.

Solution:

SR_{Max} = Value of ore − Production/Stripping cost per ton overburden
Recoverable copper per ton ore = 0.8% × 88% × 1000 = 7.04 kg
At \$1.00/kg SR_{max} = (7.04 − 5.90)/0.3 = 3.8 tons waste per ton ore
At \$1.25/kg SR_{max} = (7.04 × 1.25) − 5.90/0.3 = 9.7 tons waste per ton ore
At \$1.50/kg SR_{max} = (7.04 × 1.50) − 5.90/0.3 = 15.53 tons waste per ton ore

To check that maximum stripping ratio has been reached; for \$1.50, it is possible to strip 15.53 tons waste for each ton ore.

Profit = value of ore − production costs − stripping cost
Profit = (7.04 × \$1.5) − \$5.9 − (15.53 × \$0.3) = 0

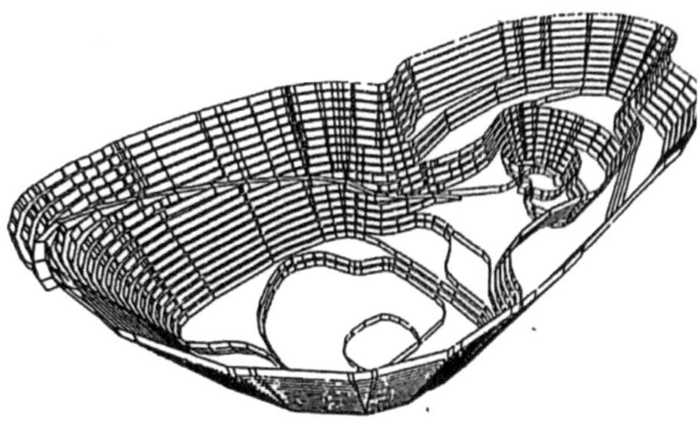

Figure 6.
A copper pit designed in this manner with varying ore grades and critical SR_{max}.

1.4 Stripping overburden methods

Overburden is a waste rock material that must be removed to expose the underlying ore body. It is preferred to extract as little overburden as possible in order to access the ore of interest, but a larger volume of waste rock is removed when the mineral deposit is deep. Most removal operations (which includes drilling, loading, blasting, and haulage) are cyclical. This is true for hard rock overburden which must be drilled and blasted first. An exception to the cyclical effect is dredging method used in hydraulic surface mining and some types of loose material mining (soil) with bucket wheel excavators. The percentage of waste rock to ore excavated is defined as the stripping ratio. Stripping ratios of 2:1 up to 4:1 are common in large mining operations. Ratios above 6:1 tend to be less economically feasible depending on the type of ore extracted. Once removed, overburden can be used for road and tailings' construction or may have a non-mining commercial value as a backfilling material. In selecting a particular stripping method and its corresponding equipment, the ultimate aim is the removal of material (waste and burden) at the least possible cost [6]. Stripping methods are classified into:

a. declining;

b. Increasing; and

c. constant.

1.4.1 Declining stripping method

In this method, each bench of ore has to be mined in sequence, and the waste in the particular bench has to be removed to the pit limit. The ore is easily accessible in the subsequent benches and the operating working space is widely available. Furthermore, all equipment usually work in the same level and so no contamination from waste blasting is left above the ore body. This method is highly productive especially at the beginning where equipment required is at minimal toward the end of the mine life. The primary disadvantage of this method is that the overall operating costs are at maximum during the initial years of operation when the maximum repayment of capital is needed and so cashflows are required to handle interest and repayment of capital (see **Figure** 7).

1.4.2 Increasing stripping method

In this method, stripping of overburden is performed as needed to uncover the ore. The working slopes of the waste faces are essentially maintained parallel to the overall pit slope angle. This method also allows for maximum profit in the initial years of operation and greatly reduces the investment risk in waste removal for ore to be mined at a future date. It is considered as a very popular method whereas mining economics or cutoff stripping ratio is likely to change in a very short time. This method is sometimes impractical because of its small spaces (narrow benches). It is available for operating a large number of equipment especially at the beginning of stripping (see **Figure 8**).

1.4.3 Constant stripping method

This method aims to remove the waste at a rate estimated by the overall stripping ratio. The working slope of the waste faces starts very shallow, but increases as mining depth increases until it equals the overall pit slope. This method has the

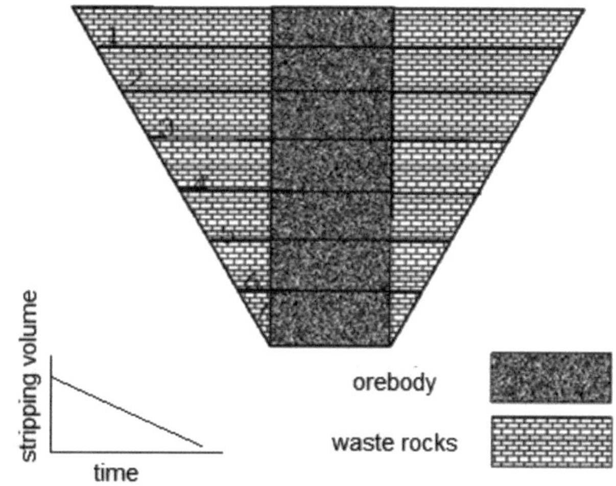

Figure 7.
Declining stripping method.

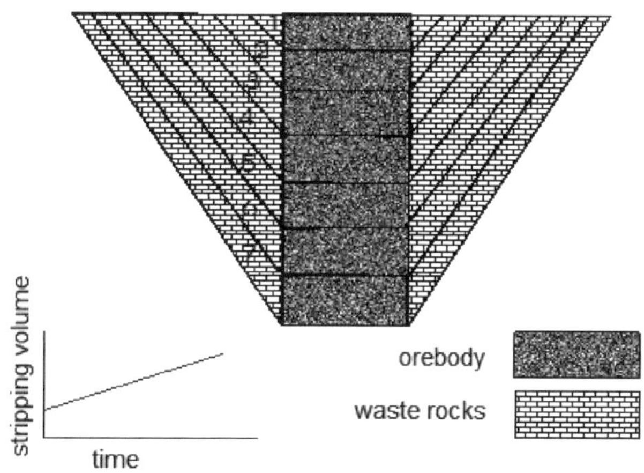

Figure 8.
Increasing stripping method.

advantage of removing the extreme conditions of the former two stripping methods outlined. Equipment fleet size and labor requirements throughout the project life are relatively constant. In this method, a good profit can be generated initially to increase cash flows. The labor and equipment fleet can be increased to maximum capacity over a period of time, and then, they can decrease gradually toward the end of the mine life. Distinct mining and stripping areas can be operated simultaneously, allowing flexibility in planning (see **Figure 9**).

1.5 Mine elements and parameters

Open pit mines are constructed of series of benches that are bisected by mine access and haulage roads angling down from the rim of the pit to the bottom. The bench height is the vertical distance between each horizontal level of the pit. The elements of a bench are illustrated in **Figures 10** and **11**, unless geological conditions dictate, otherwise all benches should have the same height. The bench height should be designed as high as possible within the limits of the size and type of the machine or equipment selected for the required production. The bench should not be so high

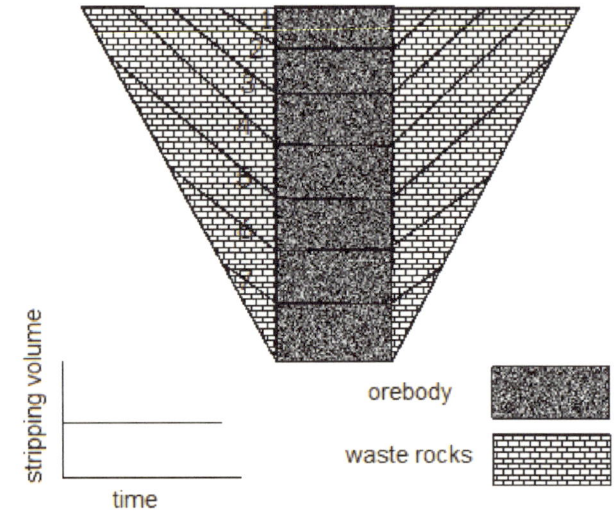

Figure 9.
Constant stripping method.

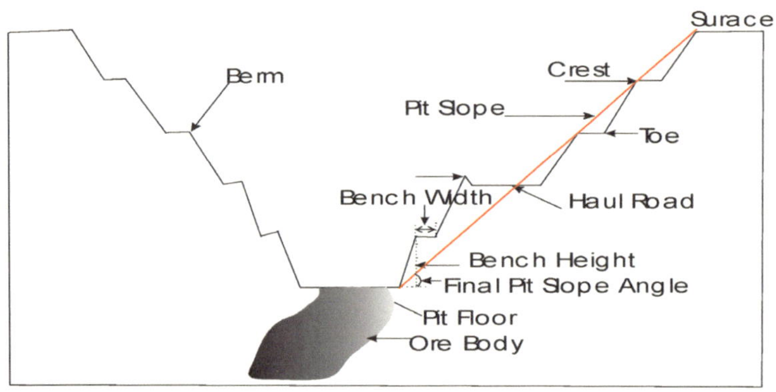

Figure 10.
Main mine elements and parameters.

that it will cause safety problems. The bench height in open pit mines will usually range from 15 m in large mines (e.g., copper) to as little as 1 m in small mines (e.g., uranium) [7]. The slope angle of the pit walls is a critical factor. If the slope angle is too steep, the pit walls may collapse. If it is too shallow, excessive waste rock may need to be removed. The pit wall has to remain stable as long as mining activity continues. The stability of the pit walls should be examined as carefully as possible. For example, rock strength, faults, joints, and fractures are key factors in the evaluation of the proper slope angle.

2. Open pit mine operations

The main economic goal in open pit mining is to remove the smallest amount of material while obtaining the greatest return on investment by processing the most marketable mineral product. The higher the grade of the ore, the greater the value received. To reduce the capital investment, an operation plan has to be developed in order to precisely dictate the way in which the ore body has to be extracted. Open pit mines vary in scale from small private enterprises processing a few hundred tons

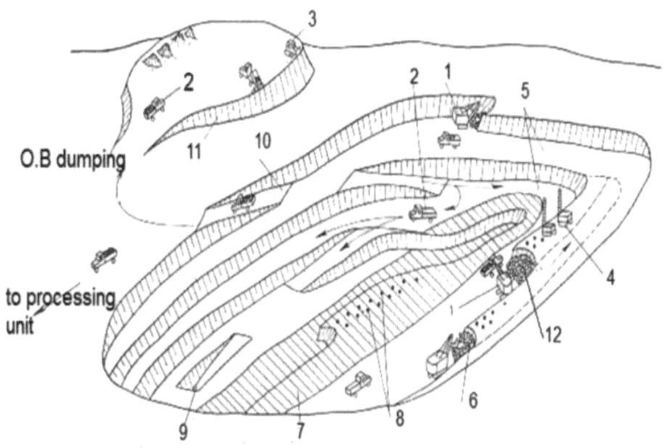

1. Excavator
2. Truck
3. Bulldozer
4. Drilling machine
5. Mining level
6. Detonated mass
7. Ore body
8. Blasting boreholes
9. Fourth level haulage Bench
10. First level access
11. O.B. Tips
12. Mining face

Figure 11.
Detailed mining operation in an open pit.

of ore a day to large companies operated by governmental and private corporations that extract more than one million tons of material a day. The largest mining operations can involve many square kilometers in area. The production cycle also referred to as the mine unit operation that consists of ripping and dozing, drilling, blasting, loading, and hauling (see **Figure 12**).

2.1 Ripping and dozing

Typically, bulldozer, wheel dozers, and motor graders are the most common equipment used, in which material transport distance is short and it can be pushed by a blade. The dozer has a large blade capacity and it is designed specifically for bulk material excavation, whereas the motor grader is a machine with a long blade used to create a flat surface during the grading process. These machines cannot “lift” the material, i.e., they do not have a load elevation capacity (**Figure 12**).

2.2 Drilling and blasting

The ore deposit can be mined by means of drilling and blasting in order to fracture the rock into a loadable size. Blasting parameters should be matched with mechanical machines for drilling of blast holes and charging of explosives. Blast holes are drilled in well-defined patterns, which consist of several parallel rows. In bench blasting, the normal blast hole patterns are square, rectangular, and staggered, **Figure 13**. The most effective pattern is the staggered pattern, which gives the optimum distribution of the explosive energy in the rock.

Figure 12.
Open pit operations cycle.

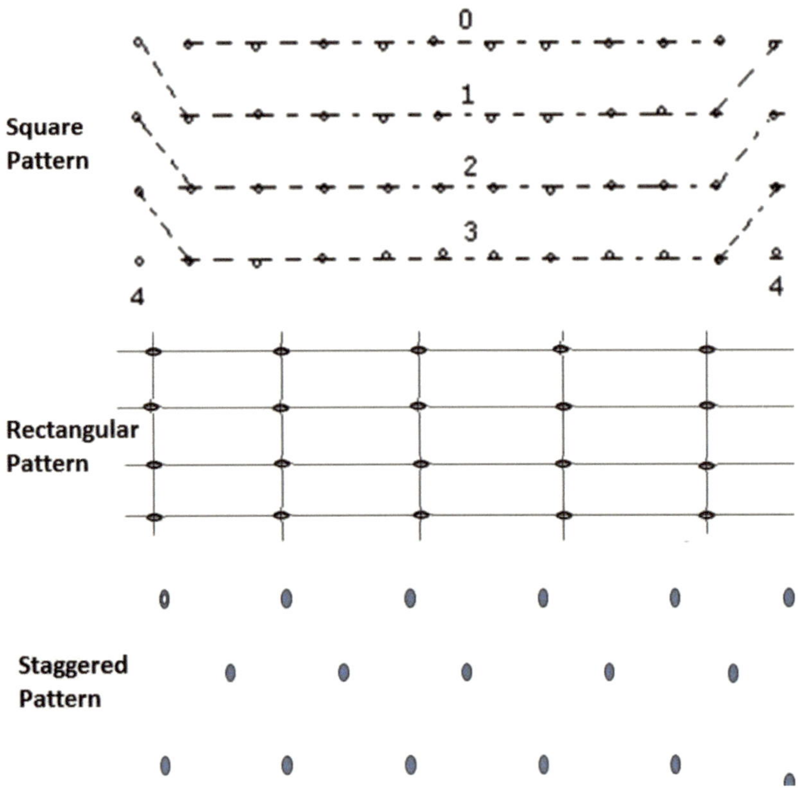

Figure 13.
Blast holes drilling patterns.

2.3 Loading and hauling

Nowadays, surface mining is conducted using shovels, front end loaders or hydraulic shovels. In open pit mining, loading equipment is matched with haul trucks that can be loaded in 3–5 cycles of the shovel. Many factors determine the preference of loading equipment. For example, with a hard digging rock, tracked shovels are more advisable. On the other hand, rubber-tyred loaders have lower capital cost and are better for loading materials that are low in volume and easy to dig. Furthermore, loaders are very mobile and well applicable for mining scenarios requiring rapid movements from one area to another. Loaders are also often used to load, haul, and dump material into crushers from blending stock piles placed near crushers by haul trucks, **Figure 14**.

Hydraulic shovels and cable shovels are common equipment used in open pit mining. Hydraulic shovels (**Figure 15**) are not chosen for digging hard rock, and cable shovels are generally available in larger sizes. Large cable shovels (**Figure 16**) with payloads of about 50 cubic meters and greater are used at mines where production exceeds 200,000 tons per day, whereas hydraulic shovels are more

Figure 14.
Loading material using front end loader.

Figure 15.
Hydraulic shovels.

Figure 16.
Cable or rope shovels.

flexible on the mine face and they enable greater operator control to selectively load from both directions (top and bottom of the mine face).

The importance of haul trucks in the history of surface mining cannot be overstated. Hand labor, wheelbarrows, horse-drawn vehicles, and ore cars were the principal means of earth-moving equipment until the early twentieth century. The advent of the internal combustion engine led to the development of the haul truck in the mining industry. Open pit mining requires a great demand for truck transport of ore and waste rock. The efficiency and greater load capacity of electrical and diesel-powered haul trucks became the preferred method for hauling in surface mining, gradually replacing rail haulage by the 1960s. Today, the average cost of a new haul truck is $3.5 million [8]. Most trucks have capacities ranging from less than 50 tons per load to 363 tons per load in large trucks such as Caterpillars 797 series load truck. Some mining companies choose to replace trucks with conveyor belt systems. For example, the Brazilian mining company "Vale" has recently replaced its mine trucks with 23 miles of conveyor belts at its iron ore mine, linking the ore deposits to the company's processing plant [9, 10].

3. Pit slope angle and stability

Slope angle is required during the early feasibility study. The degree of confidence on calculating slope angle depends upon the condition applicable. The major pit slope angle conditions can be divided into:

a. mining a shallow high-grade ore body in favorable geological and climatic conditions. Slope angles are unimportant economically and flat slopes can be used. No consideration of slope stability is required;

b. mining a variable grade ore body in reasonable geological and climatic conditions. Slope angles are important but not critical in determining economics of mining. Approximate analysis of slope stability is normally adequate; and

c. mining a low-grade ore body in unfavorable geological and climatic conditions. Slope angles are critical in terms of both economics of mining and safely of operation. Detailed geological and groundwater studies followed by comprehensive stability analysis are usually required.

During the pre-production period, the operating slopes should, however, be as steep as possible. The working slope can then be flattened until they reach the outer surface intercepts. The horizontal flow of stress through a vertical section both with and without the presence of the final pit is shown in **Figure 17**.

With the excavation of the pit, the preexisting horizontal stresses are forced to flow beneath the pit bottom. The vertical stresses are also reduced due to the removal of the rock. The rock lying between the pit outline is largely distressed. As a result of stress removal, cracks and joints can open. Cohesive and friction forces restraining the rock in place are reduced. Groundwater can more easily flow reducing the effective normal force on potential failure planes. With increasing pit depth, the extent of the stressed zone increase and the failure becomes more severe.

3.1 Types of highwall failures

There are several mechanisms by which highwall instability can occur. While we cannot expect to prevent all highwall failures, a better understanding of these

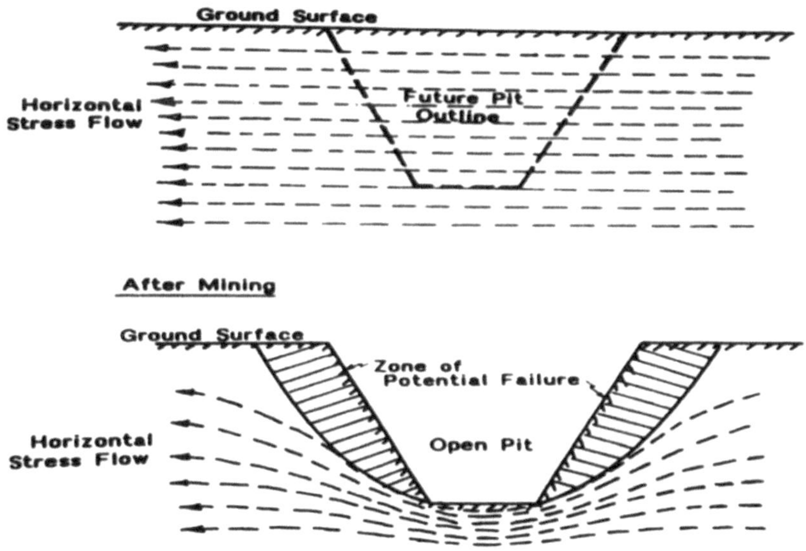

Figure 17.
Mine slope geomechanics.

mechanisms will enable us to identify potential problems before they become actual problems and to limit exposure to dangerous conditions. The most common types of failure include plane failure, wedge failure, toppling failure, and circular failure. Except for the circular failure, these usually occur along preexisting discontinuities. Example of each are the following:

3.1.1 Planar failure

This slide in **Figure 18** illustrates a typical plane failure of a highwall. Notice that the rockslide occurs along this discontinuity which daylights on the highwall and dips toward the pit. If this sliding plane does not daylight, or dips away from the pit, the slope is stable. Even if the joint daylights, in order for the slide to occur, the weight of this sliding block must exceed the frictional resistance along the discontinuity. **Figure 19** shows an example of a slope, which is plagued by large planar failures, and leads to a slide off rocks along natural, parallel, and bedding planes.

3.1.2 Wedge failure

A wedge failure occurs when two discontinuities meet and their intersecting line daylights on the slope face and dips toward the pit. If these conditions do not occur,

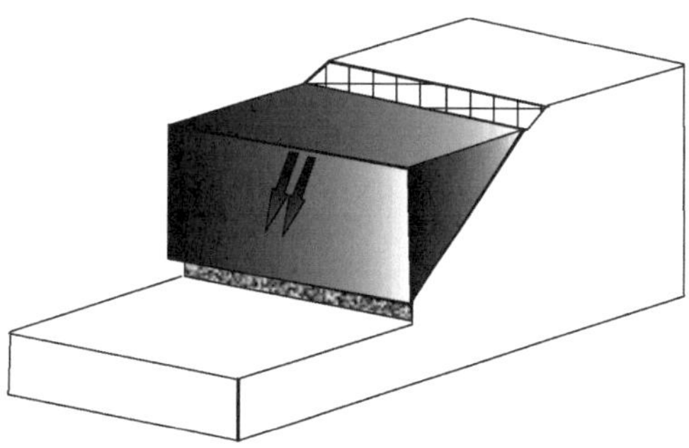

Figure 18.
Planar failure.

Figure 19.
Large planar failure.

you cannot have a wedge failure. The weight of the block also has to exceed the frictional resistance along the failure surface to have failure, **Figure 20**.

As shown in **Figure 21**, the failure can follow trends since joints tend to occur in repeating patterns. Note here the failure on the top bench, and on the next bench, should probably expect another at the next level down.

3.1.3 Toppling failure

Toppling failures look like **Figure 22**. A toppling failure can occur when the discontinuities dip very close to vertical but away from the pit. The discontinuities can be natural or they can be caused by the mining process.

If the mine progresses from left to right, there will be continuous problems, because of the way these cracks are oriented. On the other hand, if the mine goes from right to left, mine operators do not have to worry about toppling-type failure; so, decisions made during mine planning can have a profound effect on the stability

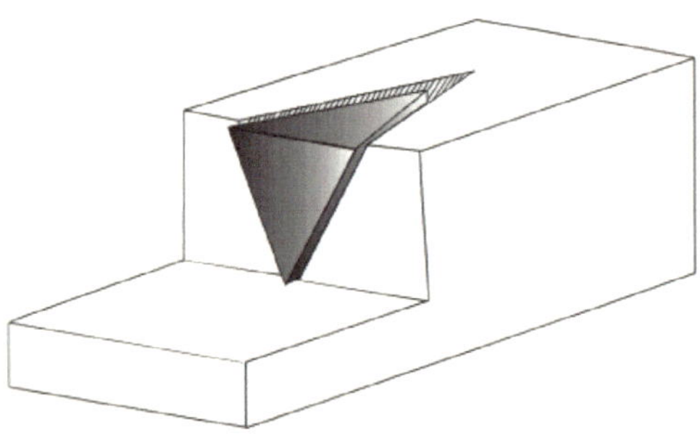

Figure 20.
Wedge failure.

Figure 21.
Trend in wedge failure.

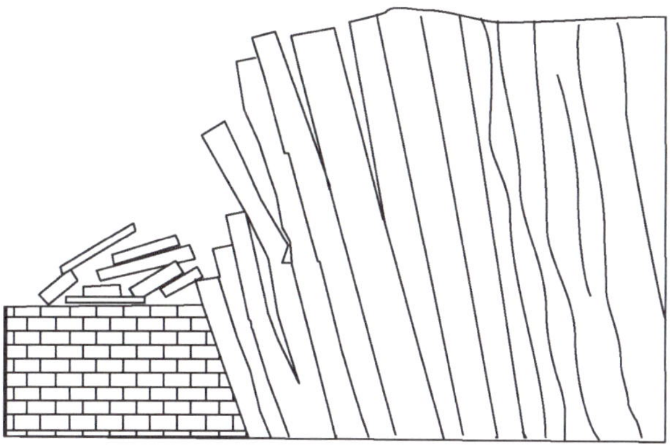

Figure 22.
Toppling failure.

Figure 23.
Toppling failure example.

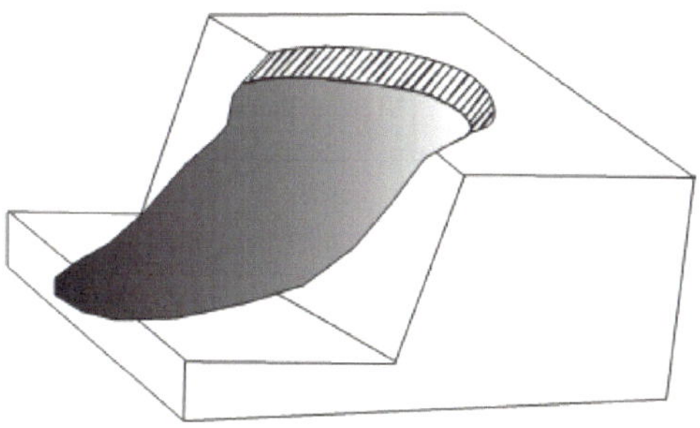

Figure 24.
Circular failure mechanism.

of the highwalls. **Figure 23** shows a picture of a toppling failure that resulted in a fatality to a blast hole drill operator.

3.1.4 Circular failure

In slopes excavated in soil or highly jointed and weathered rock mass where there are no geological structures to control the failure, the most unstable failure surface is approximately a circular arc. This circular failure surface (**Figures 24** and **25**) results from a process of localization of deformations. It is an arch type of landslides; however, the specific shape of this failure surface and the associated failure mechanism cannot be generalized [11].

Figure 25.
Circular slope failure at the open pit of the Bingham Canyon Mine in Utah.

4. Mine closure and reclamation

In general, mining has a significant negative impact on environment. Due to its nature, it leads to severe degradation of the landscape. Many factors such as drainage, air, soil and water quality, noise levels, ground vibrations, human health, and habitation are mostly affected by mining activities. When the extraction of mine reserve is over, the distorted landscape has to be reclaimed in order to reduce the damaging effects of open pit mining and bring back the landscape and its surroundings, see **Figure 26**.

Figure 26.
Open pit mining before and after reclamation.

Land use plan at the end of mining has to be set in order to determine what the mine site will look like and how the lands will be used after the mine is closed and fully reclaimed. The mine must operate and close such that the land and water in and around the mine site are less disturbed and environmentally safe and sound like original. It is the responsibility of the mining company to pay for reclamation and closure costs. To ensure that funds are available for closure, the mining company will normally be required to post a financial security (a reclamation bond) before mining starts.

"Progressive reclamation" is usually part of the overall closure plan. Progressive reclamation means that once a part of the mine site is no longer needed, it will be reclaimed rather than waiting for all aspects of operation to cease. For example, waste rock piles will be reclaimed as soon as they have reached their permitted size.

The general rehabilitation goals require rehabilitation of areas disturbed by mining to result in sites that are safe to humans and wildlife, nonpolluting, stable, and able to sustain an agreed postmining land use. The process of reclamation normally involves the following steps [12–14]:

a. *Recontouring*: the ground is re-sloped and contoured to a profile that will be stable and that provides proper drainage, facilitates the growth of vegetation, and provides various habitats for wildlife.

b. *Capping with a growth medium*: waste rock piles and other areas of the mine site will need to be covered with a soil material that is suitable for the growth of plants.

c. *Seeding and fertilizing*: this usually takes place over many years. Fast growing grasses may be planted in order to stabilize the soil followed by shrubs and trees depending on the end use plan.

d. *Monitoring*: plants in areas that are to be used for grazing will be tested to ensure they contain acceptable levels of metals and other possible contaminants.

5. Variants of surface mining methods: strip mining, mountainous mining, and artisan mining

Variants of open pit mining are limited to a number of other surface mining methods, which include strip mining, high wall mining, and quarrying. Strip (open cast) mining is used extensively for the surface mining of important commodities such as coal and phosphate ores. Casting is the process of excavation and dumping into a final location. This type of mining involves removing the overburden and extracting the valuable mineral deposit. Strip mining is applicable to shallow, flat-lying deposits [15]. It is a method that is generally applied on a large scale with low mining costs and high productivity and that has minimum land degradation [16, 17]. In Jordan, strip is used for the extraction of oil shale and phosphate ores. These mines are located at the central and southern parts of the country (e.g., **Figure 27**).

Strip mining differs from open pit in that the overburden is not transported to waste dumps but cast directly into adjacent mind-out panels, i.e., reclamation is contemporaneous with extraction. These mines often occupy a large area of land for ore excavation and overburden disposal. Strips are large rectangular parallel pits that extend to more than a mile in length [18]. After the removal of vegetation and topsoil, the mining begins with an initial rectangular box cut. The dragline is used

Figure 27.
Oil shale extraction project at Attarat Oil Shale mine, Jordan.

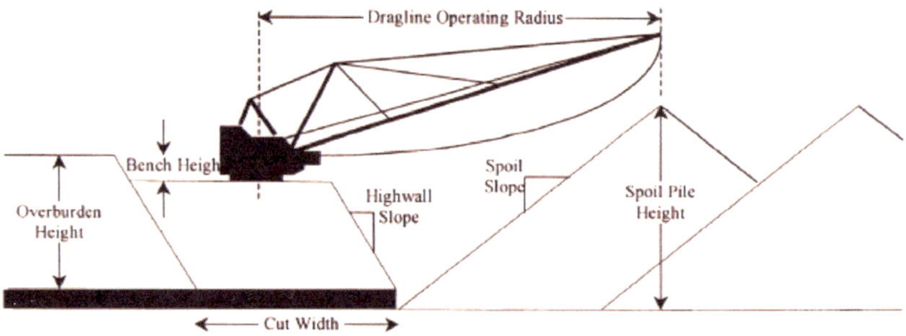

Figure 28.
Typical dragline operation [20].

Figure 29.
During loading A1 phosphate layer at Al-Shidiyah phosphate mines.

for overburden removal. As the overburden is removed from one portion of a mineral deposit, it is used to fill in the trench left by the previous removal [19]. The backfilled area is then replanted during the reclamation process.

Figure 28 shows typical dragline operation. Stripping process continues along parallel strips. Where the deposit becomes thinner, or dipping more below the surface, or in the case of dramatic increase in the stripping ratio, the mining operation must be ceased [19]. Shovel-truck system is currently adapted for extracting phosphate ore in several phosphate mines in Jordan (**Figure 29**); especially in Al-Shidiyah, Al-Abiad, and Al-Hasa mines. Since shovel truck removal of

Figure 30.
Dragline removes overburden at Al-Shidiyah phosphate mine, Jordan.

Figure 31.
Overburden removal at Attarat Oil shale mine, Jordan.

overburden generally costs at least three times as much as dragline stripping, the dragline is currently implemented for removing overburden from phosphate ore in those mines (**Figure 30**). On the other hand, shovel truck removal of overburden is currently used in Attarat oil shale mine (**Figure 31**).

In the mountainous and hilly terrains, contour mining is applied. It is also known as mountaintop mining. The mining of flat deposits in these areas follows the contour around the hill and into the hillside up to the economic limits. The extraction becomes difficult with inclination and depth increase. The top of a mountain is removed to recover the ore contained in the mountain that resulted in huge quantity of excess spoil that is placed in valleys that affected the streams flowing within these valleys [21].

Artisanal mining is a small scale mining method, which includes enterprises or individuals that employ workers in developing countries who are poor and have few other options for supporting their families and who usually use manually intensive methods for mining (e.g., panning in case of gold). Artisanal miners use elementary techniques for mineral extraction and often operate under hazardous, labor-intensive, highly disorganized, and illegal conditions [20, 22].

Author details

Awwad H. Altiti, Rami O. Alrawashdeh* and Hani M. Alnawafleh
Mining Engineering Program, College of Engineering, Al-Hussein Bin Talal University, Ma'an, Jordan

*Address all correspondence to: r_rawash@yahoo.com.au

References

[1] Wetherelt A, Peter K, Wielen V. Introduction to open pit mining. In: SME Mining Engineering Handbook. 2nd ed. Colorado: Society for Mining, Metallurgy, and Exploration Inc; 2011. p. 2161

[2] Tomakov P, Naumov I. Technology Mechanization and Organization of Surface Mining. 2nd ed. NEDRA: Moscow; 1986. p. 312

[3] Bullivant D. Current surface mining techniques. Journal for the Transportation of Materials in Bulk: Bulk Solids Handling. 1987;7:827-833

[4] Hustrulid W, Kuchta M, Martin R. Open Pit Mine Planning and Design—Volume 1: Fundamental. 3rd ed. London: Taylor and Francis; 2013. p. 995

[5] Introductory HH. Mining Engineering. 2nd ed. New York: Wiley-Interscience Publication; 2002. p. 564

[6] Oggeri C, Fenoglio T, Godio A, Vinai R. Overburden management in open pits: Options and limits in large limestone quarries. International Journal of Mining Science and Technology. 2019;**29**(2):217-228

[7] Harraz H. Mining Methods: Surface Mining Planning and Design of Open Pit Mining [Internet]. 2016. Available from: https://www.slideshare.net/hzharraz/surface-mining-planning-and-design-of-open-pit-mining [Accessed: 15 November 2019]

[8] Ascarza W. Mine Tales: Huge Haul Trucks Changed the Face of the Mining Industry. 2014. Available from: https://tucson.com [Accessed: 15 November 2019]

[9] Dessureault S. Equipment Operations Technology, University of Arizona Mining and Geological Engineering Course Notes–2005; 2015. p. 208

[10] McShane J. Metal Products Advances in Material Hauling for Mining Operations. 2020. Available from: https://www.mcshanemetalproducts.com/blog-post/advances-material-hauling-mining-operations/ [Accessed: 10 November 2019]

[11] Fleurisson J. Slope design and implementation in open pit mines: Geological and geomechanical approach. Procedia Engineering. 2011; **46**:27-38

[12] Stevens R. Mineral Exploration and Mining Essentials. British Colombia: Pakawau GeoManagement Inc./British Columbia Institute of Technology; 2014. p. 317

[13] Kuter N. Reclamation of Degraded Landscapes Due to Opencast Mining. Cankiri, Turkey: Department of Landscape Architecture, Faculty of Forestry, Cankiri Karatekin University; 2013. DOI: 10.5722/55796

[14] Liu H. Strategic Planning for Dragline Excavation Sequencing [PhD thesis]. Australia: The University of Queensland; 2018. Available from: https://espace.library.uq.edu.au/ [Accessed: 18 April 2020]

[15] Haldar S. Mineral Exploration: Principles and Applications. 2nd ed. Oxford: Elsevier; 2018. p. 378. ISBN: 9780128140222

[16] Brown I. Strip mining. Ch. 10.11. In: Darling P, editor. SME Mining Engineering Handbook. Vol. 1. Colorado: SME; 2011. pp. 1031-1046

[17] Harraz H. Mining Methods: Part I-Surface Mining. 2010. Available from: https://www.researchgate.net/publication/301824314_Mining_Methods_Part_I-Surface_mining. [Accessed: 15 December 2019]

[18] Tatiya RR. Surface and Underground Excavations—Methods, Techniques and Equipment. London: A. A. Balkema Publishers, a member of Taylor & Francis Group plc.; 2005. ISBN: 90 5809 627 0

[19] Liu H. Strategic Planning for Dragline Excavation Sequencing [PhD thesis]. Australia: The University of Queensland; 2018. [Unpublished]

[20] Erdem B, Duran Z, Çelebi N. A model for direct dragline casting in a dipping coal-seam. The Journal of the South African Institute of Mining and Metallurgy. 2004:9-16

[21] Copeland C. Mountaintop Mining: Background on Current Controversies. Congressional Research Service.7-5700; 2015. Available from: www.crs.gov. RS21421

[22] Hinton J, Veiga M, Beinhoff C. Women and artisanal mining: Gender roles and the road ahead. Chapter 11. In: Hilson G, editor. The Socio-Economic Impacts of Artisanal and Small-Scale Mining in Developing Countries. The Netherlands: August Aime Balkema, Swets Publishers; 2003

Chapter 2

Developments Made for Mechanised Extraction of Locked-Up Coal Pillars in Indian Geomining Conditions

Ashok Kumar, Dheeraj Kumar, Arun Kumar Singh, Sahendra Ram, Rakesh Kumar, Mudassar Raja and Amit Kumar Singh

Abstract

Bord and Pillar method of underground mining has been used extensively to develop Indian coal seams into pillars and galleries. This results in only 20–30% recovery of coal and rest coal remain locked up in developed pillars. Indian coalfields are famous in the world for its uniqueness and complexity of the geomining conditions which makes the extraction of the locked-up coal pillars a difficult and hazardous activity using different underground mining methods. Indian mining industry has introduced mechanisation since last 10 years to deal with the various underground rock mechanics issues in order to improve the efficiency and safety during recovery of locked-up coal pillars. But mere introduction of mechanisation did not solve all the rock mechanics problems due to requirement of indigenous design of different involved geotechnical elements for Indian geomining conditions. CSIR-CIMFR is a national research organisation engaged in improving conditions of underground coal mines. It has developed rock mechanics advances, namely, design of irregular shaped heightened rib/snook, roof bolt-based breaker-line support, warning limit of roof sagging, and cut-out distance for continuous miner-based mechanised depillaring. This chapter presents the developments made and highlights challenges to pursue future research studies for mechanised depillaring-based mass coal production from Indian underground mines.

Keywords: continuous miner, mechanised depillaring, rock mechanics issues, roof sagging, rib/snook, breaker-line support, cut-out distance

1. Introduction

Around 96% of the total coal production in India is currently being produced by opencast mining method and the contribution of underground mining is on a declining trend from 22% in 2001 to 4% in 2019. Opencast is favoured due to availability of reserves at shallow depth of cover and heavy earth moving mechanised technologies over underground as the former has rock mechanics issues as only slope/dump stability. However, opencast mining method has limitations of depth and associated environmental concerns. Underground mining is a way forward towards clean coal production technology and sustainable development. Depletion of coal at shallow depth is paving the way towards underground mining.

Indian coal mining industry had rampantly developed a number of coal seams using Bord and Pillar (B&P) mining method on square/rectangular pillars and galleries with around 20–30% coal recovery as per Regulation 111 of the Coal Mines Regulation [1]. Development of coal seam using B&P mining method requires less technical knowledge of rock mechanics. Depillaring of the developed coal pillars becomes challenging due to complex geomining conditions of Indian coalfields namely nature of roof, geological discontinuities and surface/subsurface structures. Conventional depillaring (CD) using drilling-blasting faced issues of goaf encroachment, high induced stresses and failure of underground structures. Coal producing industries started looking for suitable mass coal producing underground technologies to meet the desired coal production. Longwall mining method was introduced long back in India during 1970s but did not get success due to the direct application of foreign technology in Indian complex geomining conditions without any *in-situ* field investigation.

Continuous miner (CM) based mechanised depillaring (MD) has been introduced as a mass coal producing technology to extract the standing coal pillars. It has proved to be successful in India and CM has been deployed in a number of Indian coal mines and many more are yet to come. It has gained the faith of industry by proving its potential of safety and production. Reason of success of CM based MD in Indian coalfields are indigenous design of different geotechnical elements like irregular shaped heightened rib/snook, roof bolts-based breaker-line as goaf edge support, warning limit of roof sagging in geotechnical instrument and cut-out distance in different geomining conditions. Average daily production from a CM face is around 2000 t which is around 10 times of the daily production from a CD face using drilling-blasting. Success of any underground mining method depends upon the performance of underground structures under extreme difficult high induced stress condition. **Table 1** shows the details of MD during development and depillaring in Indian coalfields using CM.

Name of mine	Depth	Pillar size (m × m)	Gallery width (m)	Immediate roof	Manner of pillar extraction	Snook size (m^2)	Roof sagging limit in AWTT (mm)	Cut-out distance (m)
A	50–163	33 × 33	6.6	Sandy shale	Split and slice	26	5 mm as both warning and withdrawal limits	14 m in split and 10 m in slice
G	160–325	35 × 36	6.0	Sandstone	Split and slice	102	5 mm as warning and 10 mm as withdrawal limits	15 m in split and 12 m in slice
P	50–120	18.5 × 19.5	6.6	Shale	Christmas tree	22	5 mm as warning and 8 mm as withdrawal limits	12 m in split and 9.5 m in slice
V	50–100	18.5 × 19.5	6.6	Shale	Christmas tree	22	5 mm as warning and 8 mm as withdrawal limits	12 m in split and 9.5 m in slice

Table 1.
Details of mechanised depillaring operations at different Indian coalfields.

2. Rock mechanics challenges in mechanised depillaring

Underground mining in India has not boomed to that extent due to the less rock mechanics advances in B&P mining method. Depillaring continues to be one of the most challenging and hazardous activities in underground coal mining due to different accidents by roof/side falls and poor performance of structures. It is the main stage of production with around 60–80% of coal recovery. MD was first introduced in 2003 at Anjan Hill Mine of Chirimiri Area of South Eastern Coalfields Limited a subsidiary of Coal India Limited. Irregular shaped rib/snook created during pillar extraction and resin grouted roof-bolts are used as goaf edge support for the first time in India. Due to non-availability of empirical formulation to design such rib/snook and roof bolts-based goaf edge support, MD achieved mixed results in Indian coalfields. Also, the time interval between flashing of light in auto warning tell-tale instrument and roof fall was recorded to fix a warning threshold limit of roof sagging. It was successful at Anjan Hill Mine and MD was further introduced at a number of Indian coal mines to extract standing coal pillars and virgin coal seams.

It was found that the resin grouted roof-bolts as breaker line support (RBBLS) installed directly at the goaf edge did not work effectively and the roof fall extended inside the working and caused collapse of rib/snook and burial of CM [2, 3]. A small increase in area of rib/snook by 20–40% increased the stand-up time of roof in goaf by 5–10 hours. Hanging roof is a serious problem during MD as it creates the issue of front abutment stress causing goaf encroachment and burial of CM. Safety and productivity are the main concern during the underground mining. Geotechnical investigations found that caveability of overlying strata and size of remnants are the two important factors which affect the safety and productivity. Insufficient knowledge of geological discontinuities further aggravates these issues. Rock mechanics challenges at the goaf edge during MD are very complex which needs to be addressed indigenously.

2.1 Irregular shaped heightened rib/snook

Natural supports (pillar/fender/rib/snook/stook) are an important element for the success of MD. Size of pillar remnant is critical for the regular caving of overlying hanging strata in goaf during MD. Different countries used different nomenclature for the remnants like snook/stook 'x'/final stump/rib/narrow fender (**Figure 1**). Risk of sudden major roof falls is reduced by leaving a proper sized rib/snook against the goaf. It acts like a barrier between the slicing operation and goaf. Pillar is split into two equal halves and each half is called fender/stook. After splitting, one fender keeps supporting the slicing operation in another reduced sized fender. Rib/snook is remains of fender left to temporarily support the cantilevering/beaming roof to permit safe extraction and fall gradually after the extraction is completed. Stability and competency of fender and rib/snook is important for the maximum possible extraction during MD. Natural supports provide more support to the roof than any artificial designed support (cog/chock/bolt/mobile breaker-line roof support). Interaction between the support (natural/artificial) and roof determines the safety and efficiency of the MD.

During the retreat rib/snook created are further reduced judiciously for regular caving of roof in the goaf which involves dangerous risk of accident. Narrow rib/snook crushes easily compared to wider snook and size of snook decides the fall area, pattern and filling. Massive/strong overlying strata are more affected by the rib/snook size compared to weak/laminated strata. The practice of leaving too large and too many rib/snook in the goaf over supports the roof cantilever/beam resulting

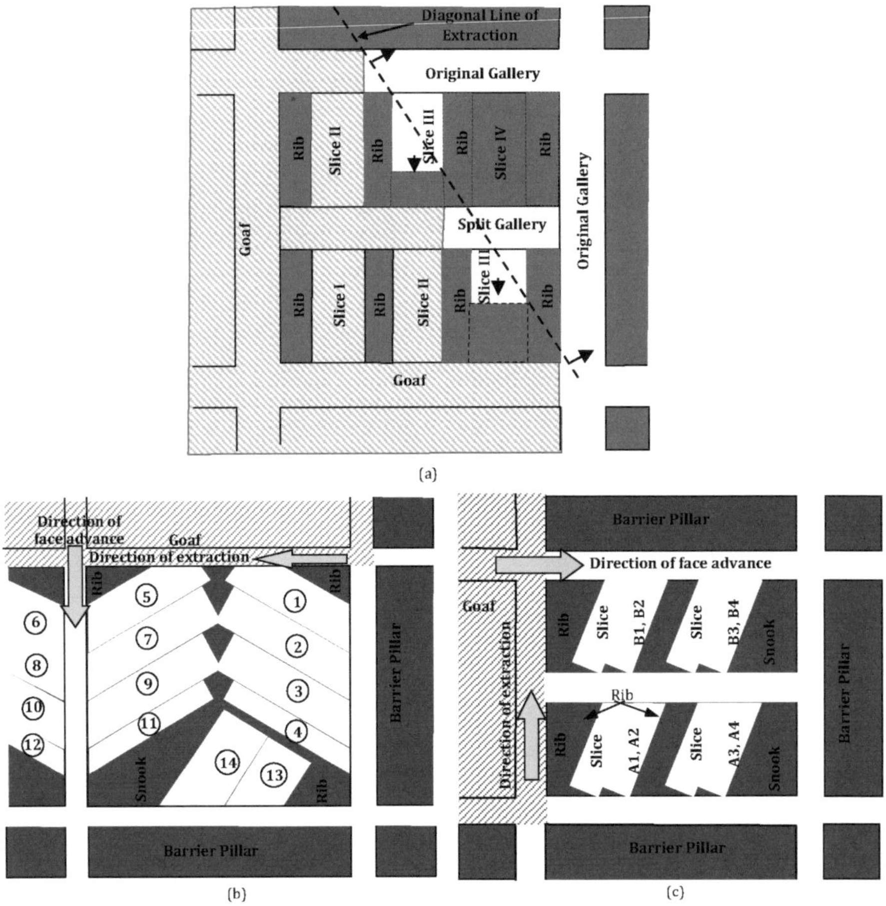

Figure 1.
Different sizes of rib/snook created during the conventional and mechanised depillaring (modified after Singh et al. [4]). (a) Manner of extraction and regular shaped rib during conventional depillaring. (b) Single pass extraction/fish-bone/Christmas tree. (c) Splitting and slicing/pocket and fender.

in increased stand-up time of roof. This results in transfer of abutment stress towards fender and solid pillars in the working. Laminated/weak strata and shallow depth cover strata have less tendency of bridging compared to strong/massive at higher depth of cover. For weaker strata even a small rib/snook acts like a solid pillar at shallow depth of cover.

Small increase in area of rib/snook may lead to increased stand-up time of roof. Different shapes and sizes of rib/snook created during the three different manner of extraction are shown in **Figure 1**. The shape of rib/snook is often irregular in shape due to the manoeuvrability of CM and existing rectangular/square pillars. Shapes and estimation of area of such shapes of rib/snook formed during CD and MD is shown in **Figure 2**. No empirical formula is applicable or available to estimate the strength of such irregular rib/snook **Figure 2(b)** and **(c)**. Seam height, depth of cover and nature of roof strata plays important role in deciding the competency of a given size of rib/snook. A conceptual model is developed for establishing a relationship of stand-up time of different nature of roof with the different sizes of rib/snook during MD as shown in **Figure 3**.

Singh et al. [4] studied the performance of rib/snook at different underground mines practising MD at different depths and nature of roof. CM carried out the slicing operation under the shadow of created competent rib/snook. Therefore,

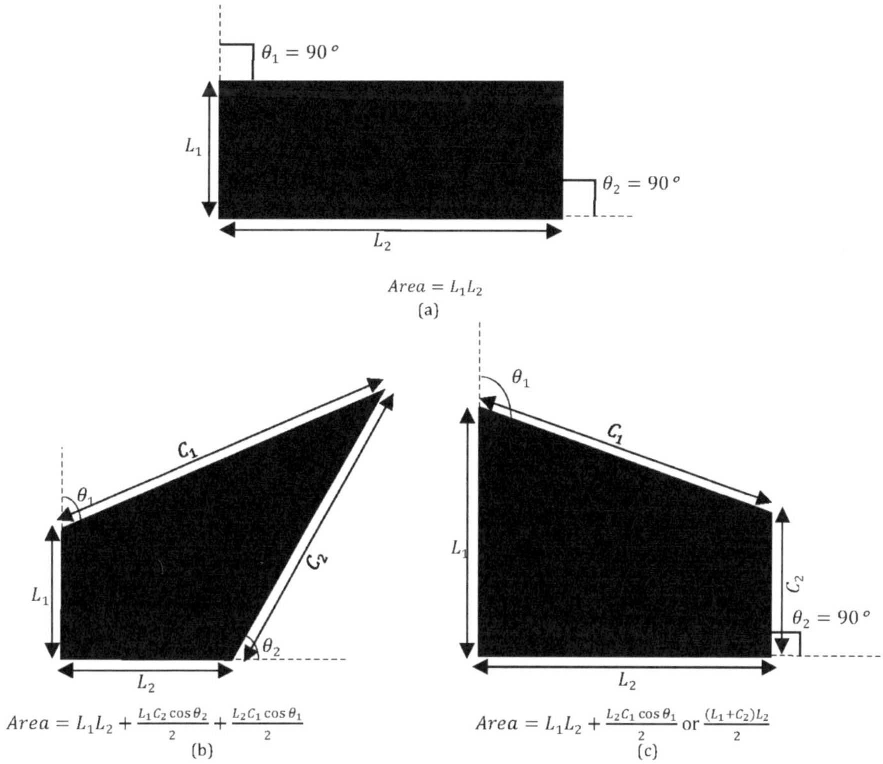

Figure 2.
Area of different shapes and sizes of rib/snook created during the conventional and mechanised depillaring. (a) Shape of snook during conventional depillaring method. (b) Shape of snook during Christmas tree/fish bone method. (c) Shape of snook during split and fender method.

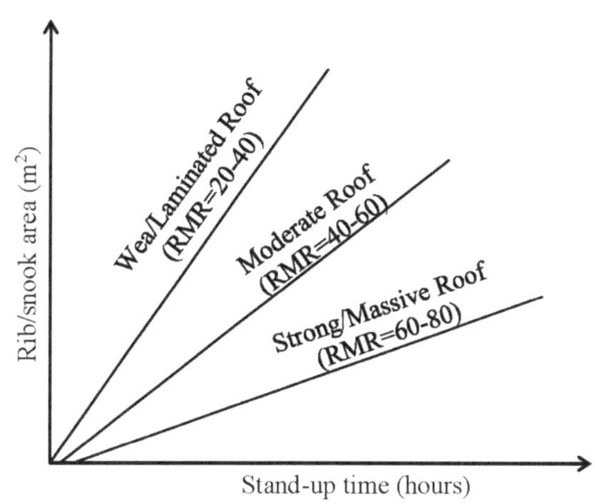

Figure 3.
Stand-up time of different nature of roof in goaf with variation with rib/snook area.

stability of rib/snook becomes a concern for the safety of men and machineries. Some cases of burial of CM occurred at few mines due to failure of incompetent rib/snook explained in Singh et al. [2]. Singh et al. [4] conducted a parametric study based on the field studies on numerical models by varying the nature of roof and depth of cover. This study found to be useful in designing a competent rib/snook during MD.

2.2 Roof bolt-based breaker-line as goaf edge support

Goaf edge during MD poses a challenging rock mechanics issue especially during the reduction of fender into rib/snook. Rib/snook formed cannot alone act against the goaf encroachment as it needs the support of breaker/hinge-line to break the bridging beam/cantilever roof. Performance of RBBLS in different mines had been monitored visually and also, using instrumented roof bolt. It was found that the position of hinge/breaker-line is affected by the nature of roof rock and size of competent fender/rib/snook/stook 'x', split gallery and out-bye intersections. Function of the hinge/breaker-line is to enhance the strength of rib/snook against caving roof and prevent the encroachment inside the working.

RBBLS forms an important geotechnical element of MD for its success. Pillar/fender at the goaf edge experienced fracturing of its sides (called spalling) which leads to shifting of position of RBBLS by 0.5–2.0 m towards the out-bye side depending upon the extent of spalling. After shifting the position, the efficiency of RBBLS enhanced (**Figure 4**). Ram et al. [5] designed the roof bolts-based breaker line as goaf edge support in Indian MD coalfields using the field and numerical simulation studies based on parametric variation of nature of roof and depth.

2.3 Warning limit of roof sagging in geotechnical instrument

Auto warning tell-tale is a geotechnical instrument which has a LED light for flashing in dark environment when the roof sagging crosses the threshold limit fixed in it. There are two important factors which decide success of AWTT in MD, namely, setting of safe roof sagging threshold limit and the fixation of anchorage point. Generally, the anchorage is fixed at a horizon of 10 m in the roof which is found to be a successful practice as the roof below it is vulnerable to failure during local fall after extraction. Roof sagging value is found to be varying in different methods of mining and factors like size of remnant, thickness and elasticity of roof affected it. Therefore, Kumar et al. [6] studied the roof sagging limit set in AWTT at different MD faces. Further, a parametric study to estimate a safe warning roof sagging limit is decided based on field studies and numerical simulation.

A typical observation by AWTT is shown in **Figure 5** indicating the time-interval between flashing and roof fall in a MD panel. Initially, due to the support by barrier pillar from three sides and solid pillar from one side, the time taken for roof fall is the maximum. As the extraction progresses away from the barrier on dip-side, the time-interval between flashing of AWTT and roof fall reduced due to the

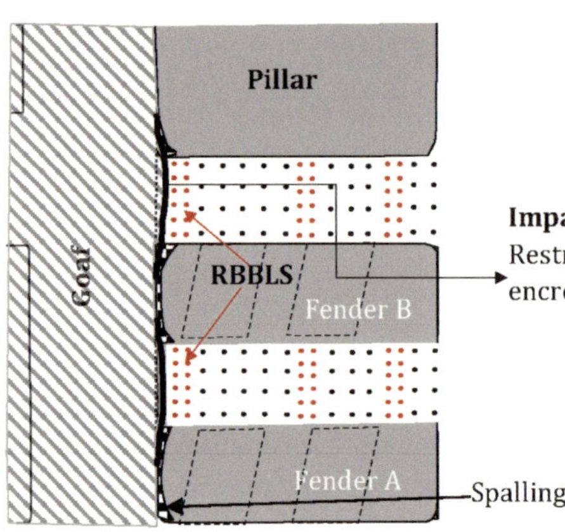

Figure 4.
Controlled caving inside the goaf after placement of an efficient breaker line supports at the goaf edge.

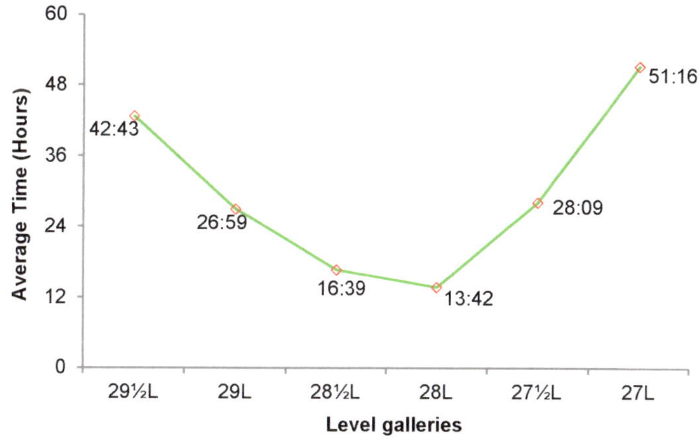

Figure 5.
Time-interval between flashing and roof fall in a MD panel measured using auto warning tell-tale.

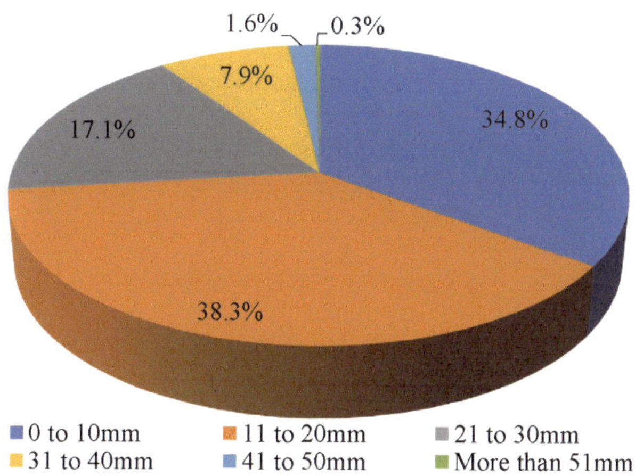

Figure 6.
Range of roof displacement observed in a panel through tell-tales.

formation of cantilever from beam. Variation in recording of roof displacement is studied in a MD panel through tell-tales (auto warning/single height/rotary). Most of the observations of roof sagging is found to be between 11 and 20 mm (**Figure 6**) as recorded from different tell-tales used in the mine.

2.4 Cut-out distance

Rock load height increases with increase in width of a gallery and found to be independent of its height. Cut-out distance defines the productivity of CM; therefore, it becomes an important geotechnical element to be designed in a panel. It is defined as the safe and stable span for a fixed width of gallery excavated by CM in a single lift without application of any applied/reactive support. Field studies are conducted at a number of Indian MD coalfields. It is found that width of the excavated gallery and nature of overlying strata are the two most influencing parameters which affects the cut-out distance to be practised in a given geo-mining conditions.

It has been also found that cut-out distance is to be kept different for a development and depillaring operations. Lesser rock mechanics issues are encountered during development activity by CM whereas depillaring involves dynamic activity where overlying strata are highly vulnerable to large induced stresses compared to development. However, it is kept to be a constant value in both development and depillaring operation for easier practice and understanding by the miners. It can be increased during development for faster preparation of the panel and reduced during depillaring for safety and productivity of the mine.

2.5 Thickness of coal seam

Thickness is a serious issue in Indian coalfields as a major amount of coal is lost due to unplanned development of a seam along roof/floor/middle horizon. A number of thick coal seams in the country are left developed along different horizons locking huge amount of coal [7]. Extraction of full thickness of a thick coal seam at a time is important for Indian underground coal mines. Earlier blasting gallery (BG) method was used by the Indian industry to extract the complete thickness of a thick coal seam in a single lift [8, 9]. But this method failed to improve the safety as well as faster and efficient recovery of coal extraction from BG panels. Height of the pillars at the goaf edge increased from the inbye side due to full extraction of the coal seam thickness during BG. Barrier pillars and pillars at the goaf edge in the panel were vulnerable to goaf encroachment and their premature collapse occurred due to strength dilution by indirect increase in height. Stability of heightened coal pillar was studied by Kumar et al. [10] using numerical simulation and established a relation between strength estimated through numerical modelling with that of CSIR-CIMFR formula, shown below.

$$S_{CMRI} = 1.28\, S_{NM} - 9.5 \quad (1)$$

where S_{NM} = pillar strength estimated through numerical modelling and S_{CMRI} = pillar strength estimated through empirical formulation.

Modification in manner of pillar extraction by CM helped in complete recovery of a 6 m thick coal at a time. CM extracted the floor coal up to 1.5–2.0 m during retreat in a slice. This created an issue of heightened irregular shaped rib/snook. Stability of such heightened rib/snook during MD was studied by Kumar et al. [7] by changing the heights of rib/snook from 3.0 to 6.0 m for a given area, nature of roof and depth. It needs to be further studied by changing the depth of cover and nature of roof with the variation in heights of rib/snook.

2.6 Issues of stress and pillar design at higher depth

Depth is a major issue for design of underground mining structures as B&P mining method is no more feasible at overburden cover greater than 400 m. Longwall is feasible at such depth of cover but indigenous design of barrier/chain/rib pillar is important. A new concept of barrierless design of longwall panel has been introduced in China. Similar concept can be used in B&P for the design of barrier pillars. Optimum design of pillars helps in maximum coal recovery and minimum wastage as left-out remnants in goaf. Worldwide available empirical formula for estimation of pillar strength does not explain the effect of depth on their strength except Sheorey [11]. Higher depth creates the case of high value of vertical in-situ stress which affects the performance of underground structures [12]. Available empirical formulation becomes redundant for the strength estimation of underground structures at higher depth. CSIR-CIMFR empirical strength formula may be used to estimate pillar strength but it did not consider the failed and stable cases of pillar at such high depth of cover [13].

Experiences of working at higher depth are important for the Indian industry as it is planning to go deeper for coal extraction in near future. Recent experiences gained by IGN, Czech Republic in collaboration with CSIR-CIMFR, are beneficial for the Indian mining industry. Underground structures at higher depth of cover needs to be designed judiciously for the maximum coal recovery by leaving less amount of coal in the goaf. Further, it would not create the issues of spontaneous heating, goaf encroachment and coal bump. Also, there is a need to design an optimum barrier pillar at higher depth of cover exceeding 300 m which crushes with retreat of working in the panel. Concept of rib/snook design in MD needs to be used for design for pillar at higher depth of cover. It should be capable to support the roof and stand stable till the extraction is over under its shadow and should fail in a controlled manner in goaf due to increase stresses. Following such design norm would help in sustainable development with maximum utilisation of resources with less wastage.

2.7 Goaf span and caveability

Different nature of overlying strata has different unsupported span for its caving. During first row of coal pillar extraction in MD, the goaf span is not sufficient for caving and therefore it is suggested to go for the maximum possible extraction due to the support from barrier and solid pillars from all the sides. Generally, roof fall is experienced when the length of the goaf span is equal to the panel length. Presence of thick difficult to cave massive competent strata does not cave even after this span due to the higher strength and thickness [14]. Hanging of such strata creates issues of goaf encroachment, over riding of pillars and sometimes air blast and coal/rock bump.

Different techniques have been used to deal with such strata during MD. MD is practised in Pinoura and Vindhya mine having easily caveable roof with frequent roof fall to VK-7 and Churcha mine having massive Deccan trap/sill delaying roof fall. Bulking factor plays a major role in caving of roof and estimation of subsidence on the surface. Sometimes the difficult to cave massive strata is located after a parting in the roof. In this case the bulking of the caved material fills the void. If the roof is difficult to cave-in and present as immediate strata then it remains hanging in goaf for a longer span of exposure and remedial measures like induced blasting or small panel (non-effective width) technique is adopted. High induced stress is created due to large span of overhang in the goaf. **Figure** 7 shows the area of exposure and progressive area of fall in a MD panel.

The value of caveability index (I) decides the easiness/difficulty hanging overlying strata to cave in goaf. It was established by Sarkar and Singh [15] for characterisation of the overlying strata in Longwall mining method. It is defined as:

$$I = \frac{\sigma l^n T^{0.5}}{5} \tag{2}$$

where σ = uniaxial compressive strength in kg/cm^2, l = average length of core in cm, T = thickness of the strong bed in m, and the factor n has a value of 1.2 in the case of uniformly massive rocks with a weighted average of RQD of 80% and above. In all other cases n = 1.

This caveability index developed for Longwall mining is not applicable in case of MD as it has a number of openings around the goaf edge and left-out ribs/snooks. There is a need to develop such index for MD which would help to extract coal by CM especially under extremely difficult to cave massive strata.

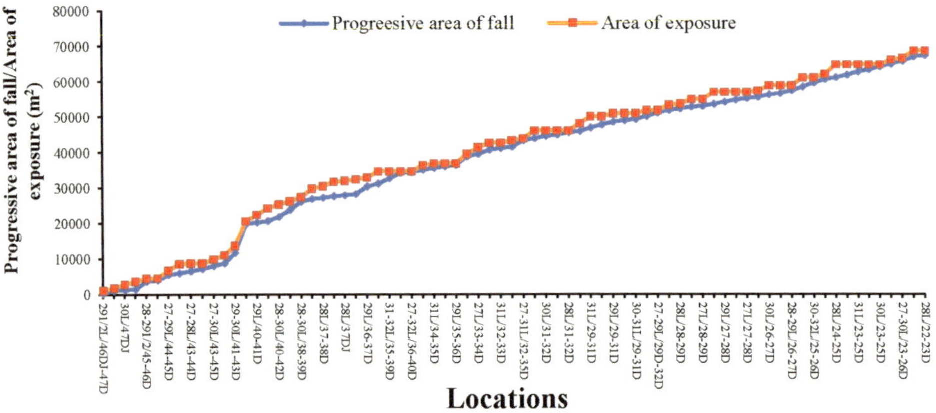

Figure 7.
Details of area of exposure and roof falls occurred in MD panel.

2.8 Geological discontinuities under influence of in-situ stress

Underground mining operation has witnessed failure of roof during widening and heightening of the developed galleries for adaptability and manoeuvrability of CM (**Figure 8**). This failure occurs due to hidden joints/slips with wash-outs, intercalation, and cross-stratification of shale and sandstone (**Figure 8**). Further, this failure was controlled by using the appropriate support system as per the results of numerical simulation and field investigations. Support becomes an important element under such disturbed nature of roof. Artificial supports in the immediate roof are installed generally after a length of 12 m (cut-out distance of CM) in a gallery width of 6 m. Geological discontinuities (hidden slips, wash-outs, cross-stratifications and intercalations of shale and sandstone) in immediate roof strata affected the advancement of drivages using CM in a panel. Freshly exposed immediate roof strata up to 1.8 m failed over the remotely operated CM. The local fall was dangerous to the drivages as it affected the safety, production and productivity of the mine. After the roof fall, cut-out distance was reduced to 4 m but the roof instability continued in the drivages. Further, support system was redesigned (increased density and length of bolts with wire mesh) to successfully control the roof instability for the reduced cut-out distance of 4 m.

Figure 8.
Wash-outs exposed in the roof of dip rise gallery reduced the cut-out distance of CM during development.

2.9 Determination of in situ strength of rock mass

It is easier to estimate the physico-mechanical properties of coal and rock in the laboratory using different rock testing equipment. These properties do not reflect the actual properties of in-situ coal/rock mass. There is available Sheorey failure criterion but it is age old and needs to be re-established for the higher depth of cover cases. Also, the strength estimated in laboratory are on a higher side and if these are considered for design of underground structures then there are likely chances of an under design which is vulnerable to fail.

Strength of rock mass is important for the stability of underground structures in rock and its realistic assessment for coal measure strata presents a unique challenge. Rock mass classifications have tried to quantify the behaviour of the rock mass. Failure criterion is helpful in prediction of strength of rock mass. But, the anisotropic and inhomogeneous behaviour of coal pillar restricts the scope of rock mass failure criteria for higher depth of cover. There is a need to revisit RMR classification system for failure criteria of intact coal measure formations at higher depth of cover.

2.10 Rock burst/coal bump

Stress concentration on underground structures results into accumulation of strain energy inside it resulting into coal bump/rock burst. Coal measure formations have the capability to store large amount of strain energy before failing. It involves the violent failure of rock/coal around an excavation causing severe injury to the miners. Indian coalfields have rare experience of dealing with coal bump/rock burst due to working under moderate nature of roof at shallow depth of cover. Some incidences of coal bump/rock burst have been experienced due to hanging of overlying strata for a longer span in goaf after pillar extraction which creates high abutment stresses over the solid pillars. It is difficult to deal with such strata as their sudden caving lead to sometimes air blast.

Deep coal mines with massive/strong roof and high stress-conditions experience coal bump/rock burst. Severity of the rock bump increases with increase in depth and stress. Instrumentation and monitoring using geotechnical instruments and micro seismic methods are helpful in understanding and prevention of such occurrences. Energy stored depends upon the physico-mechanical properties of strata. Various destressing techniques have been practised worldwide to deal with such issues.

3. Rock mechanics advances in mechanised depillaring

3.1 Design of rib/snook

Parametric study by varying the nature of roof and depth of cover was carried out in FLAC^{3D} by Singh et al. [4] to estimate the size of irregular shaped rib/snook during MD of existing square/rectangular shaped pillars. Further, height of rib was also varied [7] during extraction of complete thickness of a thick coal seam at a time using numerical simulation to estimate a stable competent size. Results of field and numerical simulation were used to estimate a competent rib/snook. Conceptual model was formulated to have a general idea about variation in size of rib/snook with depth of cover and nature of roof strata.

Area of competent sizes of ribs/snooks with variation in depth of cover and nature of the roof strata are analysed through a multivariate regression. A relationship is developed based on the analysis to estimate a competent size of the rib/snook (S), which is given as:

$$S = 0.52\,H^{0.74}\;\;R^{0.23}\;\text{m}^2 \tag{3}$$

where H = depth of cover (m) and R = CMRI-RMR.

3.2 Design of goaf edge support

Rock load height (RLH) estimated at the goaf edge using numerical models with variation in RMR and depth of cover and analysed using multivariate regression by Ram et al. [5]. Based on field studies and numerical simulation observations, relationships are developed for the design of RBBLS at three different locations around the goaf edge which are given below.

For 0 m out-bye from goaf edge

$$RLH = 11.67\,H^{0.58}\,R^{-1.14} \tag{4}$$

For 1 m out-bye from goaf edge

$$RLH = 66.32\,H^{0.31}\,R^{-1.26} \tag{5}$$

For 2 m out-bye from goaf edge

$$RLH = 115.22\,H^{0.12}\,R^{-1.20} \tag{6}$$

3.3 Prediction of roof sagging limit for roof fall

Kumar et al. [6] did a multivariate analysis of the roof sagging recorded from the numerical models with variation in thickness and elastic modulus of immediate roof, size of remnants and distance from the goaf edge. This analysis helped in derivation of an Eq. 9 to calculate the limiting roof sagging value as:

$$C = 26.63 - 0.12\text{D} - 1.12\text{E} - 0.14\text{A} + 0.23\text{T} \tag{7}$$

where C is the roof sagging observed in model (mm), D is the goaf edge distance (m), E is the elastic modulus of immediate roof (GPa), A is the size of remnants left in or around goaf edge (m^2), and T is the immediate roof thickness (m).

Taking into account the anisotropic and heterogeneous natures of rock, a safety factor of 2 is selected for fixation of the sagging value for a warning limit in AWTT which is given as:

$$S = 0.5\text{C} \tag{8}$$

where S is the warning value of roof sagging (mm) to be fixed in an AWTT.

3.4 Design of cut-out distance

CM does not damage the surrounding roof like drilling-blasting during cutting of coal. Cut-out distance in field has been practised based on trial and error in field. Numerical models based on a safe and stable roof sagging value of 5 mm are used to study the cut-out distance with varying nature of roof and width of gallery [7]. Elastic constitutive model is used to study the cut-out distance based on field studies in FLAC^{3D} by fixing the allowable range of roof sagging to 5 mm (**Figure 9**). Roof sagging values for a 6 m width of gallery by varying the cut-out distances are shown in **Figure 9** on numerical models. **Figure 9** also depicts that the cut-out distance can be further extended beyond 12 m during development using CM for faster extraction.

Based on the results of numerical model and field studies, a relationship is established to estimate the cut-out distance with variation in nature of roof and gallery width, which is given as:

$$S = 14.61 + 1.98\text{E} - 2.12\text{W} \tag{9}$$

where S is the length of cut-out distance (m), W = width of gallery (m), and E = elastic modulus of immediate roof (GPa).

4. Future rock mechanics issues

Apart from abovementioned issues for B&P mining method using CM based MD, there are challenges of rock mechanics in Indian coalfields at higher depth of cover for the characterisation of rock mass, response of underground structures to high in-situ stresses, design of underground structures, economics, subsidence, complete extraction of difficult coal seam at a time, failure criterion of rock mass, fixation of warning limit for stress and convergence in different geotechnical instrumentation and so forth.

Despite being the second largest producer of coal in the world, Longwall top coal caving method of mining is still not practised in Indian coalfields whereas China produces around 90% of the coal using this technology. Most of the Indian coal is being produced using opencast method which is not sustainable for longer duration

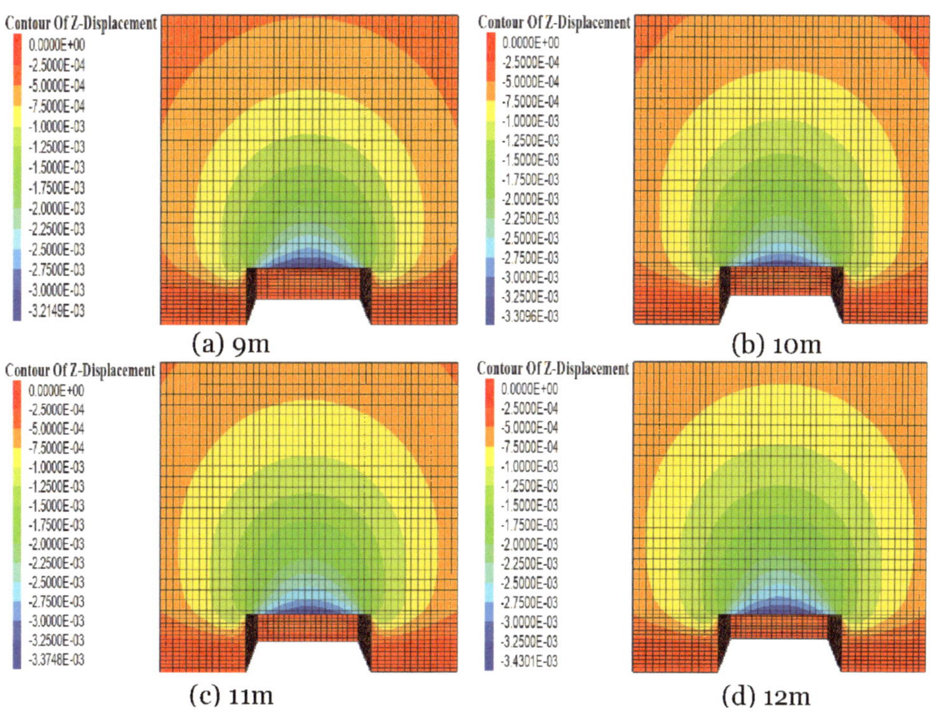

Figure 9.
Roof sagging value for different cut-out distance in FLAC^{3D}. (a) 9 m, (b) 10 m, (c) 11 m, and (d) 12 m.

due to its different limitations. Solutions to these future problems lie in carrying out R&D for each such issue on priority basis for the Indian coalfields.

5. Conclusions and recommendations

Mechanised depillaring using continuous miner technology has proven its potential in improvement of production and safety since last 10 years. A number of Indian coal mines are preferring mechanised depillaring over conventional technique to extract the locked-up coal pillars. Field study found that geo-mining conditions and design of geotechnical structures created during mechanised depillaring affect the performance of this mass coal producing technology. Rock mechanics developments in design of geo technical such as breaker-line support, rib/snook, cut-out distance/lift length and determination of roof sagging limit in instruments at the goaf edges has improved the performance of mechanised depillaring operations. However, rock mechanics issues like complete extraction of a thick coal seam, large span of overhang, caveability of difficult overlying strata, geological discontinuities, depth of working and pillar design at higher depth remains a challenge for this technology. Efforts are being made to deal with these issues in Indian coalfields as per their confrontation.

Acknowledgements

The authors are obliged to the Director of CSIR-Central Institute of Mining and Fuel Research, Dhanbad, and the Director of Indian Institute of Technology (Indian School of Mines), Dhanbad, for their permission to publish this chapter. The authors also give their due regards to the mine management of different Indian coalfields for the support during field studies. The views expressed in this chapter are that of the authors and do not necessarily reflect the opinion of their organisations.

Conflict of interest

The authors declare no conflict of interest.

Author details

Ashok Kumar[1*], Dheeraj Kumar[2], Arun Kumar Singh[1], Sahendra Ram[1], Rakesh Kumar[1], Mudassar Raja[1] and Amit Kumar Singh[1]

1 Rockmass Characterisation and Innovative Mining Methods Section, Council of Scientific and Industrial Research-Central Institute of Mining and Fuel Research (CSIR-CIMFR), Dhanbad, Jharkhand, India

2 Department of Mining Engineering, Indian Institute of Technology (Indian School of Mines), Dhanbad, Jharkhand, India

*Address all correspondence to: ashok.bhu.min09@gmail.com; ashok@cimfr.nic.in

References

[1] The Coal Mines Regulation. Published in The Gazette of India: Extraordinary issued by Ministry of Labour and Employment, Part- II-Section 3(i). 2017. pp. 160-280

[2] Singh AK, Kumar A, Kumar D, Singh R, Ram S, Kumar R, et al. Coal pillar extraction under weak roof. Mining, Metallurgy & Exploration. 2020. DOI: 10.1007/s42461-020-00277-8

[3] Kumar A, Kumar D, Verma AK, Singh AK, Ram S, Kumar R. Influence of overlying roof strata on rib design in mechanised depillaring. Journal of The Geological Society of India. 2018;**91**(3): 341-347. DOI: 10.1007/s12594-018-0860-7

[4] Singh R, Kumar A, Singh AK, Coggan J, Ram S. Rib/snook design in mechanised depillaring of rectangular/ square pillars. International Journal of Rock Mechanics and Mining Sciences. 2016;**84**:119-129. DOI: 10.1016/j.ijrmms.2016.02.008

[5] Ram S, Kumar D, Singh AK, Kumar A, Singh R. Field and numerical modelling studies for an efficient placement of roof bolts as breaker line support. International Journal of Rock Mechanics and Mining Sciences. 2017; **93**:152-162. DOI: 10.1016/j.ijrmms.2017.01.013

[6] Kumar A, Kumar D, Singh AK, Ram S, Kumar R, Gautam A, et al. Roof sagging limit in an early warning system for safe coal pillar extraction. International Journal of Rock Mechanics and Mining Sciences. 2019;**123**:104131. DOI: 10.1016/j.ijrmms.2019.104131

[7] Kumar A, Kumar D, Singh AK, Ram S, Kumar R, Singh AK, et al. Strength estimation of irregular shaped heightened rib/snook for mechanised depillaring. In: Proceedings of 8th Asian Mining Congress (AMC); 6-9 November 2019; Kolkata, India. MGMI; 2019. pp. 529-538

[8] Kumar R, Mishra AK, Singh AK, Singh AK, Ram S, Singh R. Depillaring of total thickness of a thick coal seam in a single lift using cable bolts: A case study. International Journal of Mining Science and Technology. 2015;**25**(6): 885-896

[9] Kumar R, Singh AK, Mishra AK, Singh R. Underground mining of thick coal seams. International Journal of Mining Science and Technology. 2016; **26**(2):223-233

[10] Kumar A, Kumar R, Singh AK, Ram S, Singh PK, Singh R. Numerical modelling-based pillar strength estimation for an increased height of extraction. Arabian Journal of Geosciences. 2018;**10**(18):411. DOI: 10.1007/s12517-017-3179-6

[11] Sheorey PR. Pillar strength considering *in-situ* stresses. In: Information Circular (IC), 9315. United States: Department of the Interior, Bureau of Mines; 1992. pp. 122-127

[12] Kumar A, Waclawik P, Singh R, Ram S, Korbel J. Performance of a coal pillar at deeper cover: Field and simulation studies. International Journal of Rock Mechanics and Mining Sciences. 2019;**113**:322-332. DOI: 10.1016/j.ijrmms.2018.10.006

[13] Sheorey PR, Das MN, Barat D, Prasad RK, Singh B. Coal pillar strength estimation from failed and stable cases. International Journal of Rock Mechanics and Mining Science and Geomechanics Abstracts. 1987;**24**(6):347-355

[14] Kumar A, Singh AK, Kumar D, Ram S, Kumar R, Gautam A, et al.

Caveability assessment of a hanging overlying massive Deccan trap and its effect on underground. Insights in Mining Science and Technology. 2019; **1**(3):50-60

[15] Sarkar SK, Singh B. Longwall Mining in India. Dhanbad: Sunanda Sarkar; 1985

Chapter 3

Ecofriendly Hill Mining by Tunneling Method

Rama Dhar Dwivedi and Abhay Kumar Soni

Abstract

Mostly, hills are mined by 'Strip mining' i.e. removing the hills from top. This conventional approach destroys the landscape and defaces the beauty of the hill. Besides, a large amount of dust generated at source disturbs the villagers and nearby human settlements during the excavation operation or related activities. To eliminate this, and remove the 'out yard dumping of material', except at initial stage i.e. during developmental phase, if tunneling methods of civil construction work is applied, 'the conventional hill mining' can be turned into an eco-friendly hill mining with very little planning efforts. This chapter highlights the abovementioned aspects of 'hill mining' covering overviews about the 'hill mining by tunneling method'. In this technique, the extraction of mineral deposits is done by driving tunnels at the bottom (or other accessible higher level of the hills) and combining it with cross-cuts and adits, to protect the green cover and the serene hill environment. A case study of limestone mining in hilly Meghalaya region of India forms a part of the description where its feasibility exists.

Keywords: hill mining, strip mining, tunneling in hills, mining of minerals in India

1. Introduction

Hills are being mined since long, ever since man discovered the use of metals and valuable stones. The mineral resources (called mineral deposits) do exist both above ground level and below ground level. The hills have been mainly targeted because the winning of minerals above the ground is easier as compared to the mineral deposits found at depth. For instance, a broken hill lode in South Australia, one of the largest lead-zinc lode, ever discovered, is being mined for its mineral content in hills. Similarly, mining in hills is carried out for commercial minerals like iron ore (*Kudremukh, Karnataka*, India); bauxite (*Kollimalai hill deposits of Nilgiris* in Tamil Nadu India); base metal (*Arravallis in Rajasthan*, India); Magnesite of Indian Himalayas; useful stones such as granites, slates, marble, sandstone, etc. Thus, ubiquitous 'hill mining' was existing in the past, present and will continue to remain in future as well, wherever these deposits are made available by nature for mankind.

In the context of 'hill mining', two aspects matter significantly, and they are - 'scientific extraction' and 'environmental protection'. With judicious planning and serious efforts, *the conventional approach* of mineral extraction (mining) in hills can be turned into an eco-friendly hill mining. Fragile/serene hill environment can be protected by adopting 'best practices' as applied in mining.

The characteristic features of hilly topography and the typical conditions encountered in such terrain have to be taken into account for achieving the desired results. Because favorable conditions do exist, a combination and integration of civil and mining engineering knowledge have been done and 'tunneling method' is evolved as an engineering field application. Various key aspects of hill mining covering overviews on the environment are also highlighted and described in the chapter. A selective case record of limestone mining in the Meghalaya state of India forms the part of the description for such type areas, the reason being its feasibility.

Conceptualization of tunneling method, though not new, came into our mind around the la*te* nineties *(≈ 1996 to 1998) when* limestone mining by underground methods was tried in an experimental adit in Himachal Pradesh, India**.** This site was located in the hilly Himalayas and the framed experimentation was found quite effective to protect the sensitive and fragile Himalayan environment [1]. In this way, integration of our knowledge about conventional 'underground method of mining 'and tunnel excavation work, prominent in hydel power projects of Himalaya, have to lead to the development of, *ecofriendly tunneling method*, considering the hill topography (or hill areas) as our concentration point.

2. Hill mining by tunneling method

Mostly, hills are excavated by 'stripping method', which consists of removing the top hill cover and moving downward (chopping down) in steps. When hills are mined for mineral extraction, the methodology of mining is termed as **'Strip Mining'**. This method is a conventional method and the extraction is carried out by construction of berms (or benches) to reach the deposit, and excavating mineral by digging either manually or mechanically. The conventional approach destroys the landscape and defaces beauty of the hill (**Figure 1**). Also, a large amount of dust generated at the source disturbs the surroundings, villagers and nearby human settlements during ongoing excavation operation or related ancillary activities.

Figure 1.
Defacing of hills due to illegal quarrying is a typical sight in many hilly areas (severe environmental impacts to natural hill landforms).

To eliminate and overcome this, 'tunneling methods', as used in civil construction work, maybe the alternative that can be applied.

Tunneling method involves number of tunnels driven either in the country rocks and/or in the ore/mineral formation itself. The size of tunnels is chosen based on the thickness and compressive strength of the orebody and host rocks encountered.

It is well known that the economics of mineral extraction depends on the adopted mining method and its market value. Location, orientation, size and strength of ore deposit are the prime influential parameters to choose a mining method. However, nowadays environmental conditions are forcing the decision-makers to indulge in the activities which are ecofriendly or at least it should not harm the flora and fauna of the landscape. In this context, going underground without disturbing the natural surface features is appreciated for mining the minerals from the ground. In addition to this, the excavated *underground space* is preferred to be re-utilized either as a waste material refill or as a valuable space for miscellaneous purposes, to avoid subsidence at the ground surface above the mined-out area. Thus, rehabilitation of the excavated space is a value addition by public or industry. In such condition, the underground mined out areas need to be well supported so that these can stand for several coming years and thus accrues to the cost of mining.

If the mineral deposit in a hill is found feasible for mining, it shall be mined using 'tunneling method ', which would involve the following steps.

A. Geological and geotechnical investigations for planning

To extract mineral, location, thickness and alignment of mineral/orebody should be known as it influences the preparation of actual excavation plan (mining plan), to be implemented into practice. Overburden or rock cover should be known because it gives an idea about induced stresses around excavation built underground. Here, the size and diameter of the underground opening play an important role in the stability. Further, rock mass properties, information of water condition and physico-mechanical properties of rock mass decide the requirement of support needed for the excavated area (i.e.tunnel walls) for the required life span.

Both, geological and geotechnical investigations help in the planning of mine and execution of various unit operations that lead to the extraction of mineral from the earth. These investigations reveal the following information about the mineral deposit/ore body and their properties.

i. Thickness and alignment:

Thickness indicates the volume of the orebody and determines the economy for mining activity and alignment or orientation is a deciding parameter for mining methodology. It is estimated with the help of core-log details obtained from various boreholes.

ii. Overburden or rock cover above the orebody:

This is the rock cover thickness above the orebody. It helps in the estimation of vertical in-situ stress, if any, at an underground place of workings.

iii. Rock joint properties:

Number and properties of rock joints together with the strength of rock mass helps to know its behavior when subjected to induced stresses during excavation activities below ground. Rock joints present also helps in knowing the water permeability of the strata.

iv. Location of the water table, if present:

Depth of water table provides information about water head to be considered while designing supports for roof or walls of the excavated area. In addition to this, it also gives an idea about the expected quantum of water inside the workings below ground.

v. Physico-mechanical properties:

Physical properties, like permeability (K) and specific gravity (γ) help the support designer in the estimation of the rate of water inrush and value of vertical in-situ stress respectively. Whereas, mechanical properties like Uniaxial compressive strength (σc), modulus of elasticity (E), Poisson's ratio (ν), cohesive strength (c), and angle of internal friction (ϕ) play a vital role in determining deformational behavior of rock mass (country rock and orebody) while excavation goes below ground during actual mining.

B. Preparation of a mining plan.

It includes preparation of geotechnical baseline reports, structural design report and drawings, for the approach roads, excavation sequence of tunnels and cross-cuts, applicable supports and drainage plan. As the reports and drawings are prepared based on geological and geotechnical data explored from the surface, the reports and drawings are revised when actual geology and rock types are encountered i.e. during going underground.

C. Design of tunnels and cross-cuts.

Size of tunnels is decided based on the rock mass quality, thickness of the orebody and rock cover. The rock mass is characterized and its quality is assessed using Barton's Q-system, rock mass rating (RMR) system or geological strength index (GSI) system. Rock mass with RMR value greater than or equal to 80 has high *standup time*. For example, 10 m of unsupported tunnel span can withstand up to about 70 months in a rock mass with RMR of 80 (Q = 55). Due to this reason, tunnel excavated in such quality rock mass does not require any artificial supports except spot bolting. On the other hand, the requirement of supports increases with a decrease in the quality of rock mass. Based on rock quality (Q-value), the quantum of supports required for tunnels of 5 m and 10 m diameter have been listed in **Table 1**, which have been obtained from the Barton's Q-chart as given in **Figure 2** [2].

Excavation support ratio (ESR) is the weightage assigned to the type of structure based on their importance. The more important structure is assigned with a lower value of ESR (**Table 2**). Value of ESR for temporary mining opening has been assigned as 3–5. Average ESR value (3) has been considered for calculation of supports in **Table 1** for the tunnels or cross-cuts, which are temporary and shall be backfilled after mining of the mineral/ore. On the other hand, the tunnels or cross-cuts, which shall be retained as rehabilitated underground space are permanent and have been assigned with ESR value of 1.6.

This is significant to note that an arched crown of the tunnel distributes the induced stresses around the tunnel boundary in a better way. Required supports

Rock quality (Q)	Tunnel span (D) =5 m				Tunnel span (D) =10 m			
	ESR = 1.6		ESR = 4		ESR = 1.6		ESR = 4	
	D/ESR	Supports	D/ESR	Supports	D/ESR	Supports	D/ESR	Supports
0.4	3.125	Nil	1.25	Nil	6.25	L = 2.5 m, S = 1.5 m, Sc = 9 cm	2.5	Nil
1	3.125	Nil	1.25	Nil	6.25	L = 2.5 m, S = 1.7 m, Sc = 8 cm	2.5	Nil
4	3.125	Nil	1.25	Nil	6.25	L = 2.5 m, S = 2.1 m, Sc = 6 cm	2.5	Nil
10	3.125	Nil	1.25	Nil	6.25	L = 2.5 m, S = 2 m	2.5	Nil
40	3.125	Nil	1.25	Nil	6.25	Nil	2.5	Nil
50	3.125	Nil	1.25	Nil	6.25	Nil	2.5	Nil

Notations: ESR-Excavation support ratio; L-Bolt length; S-spacing of bolt length; Sc-Reinforced shotcrete.

Table 1.
Requirement of supports according to rock mass quality and span or diameter of the tunnel.

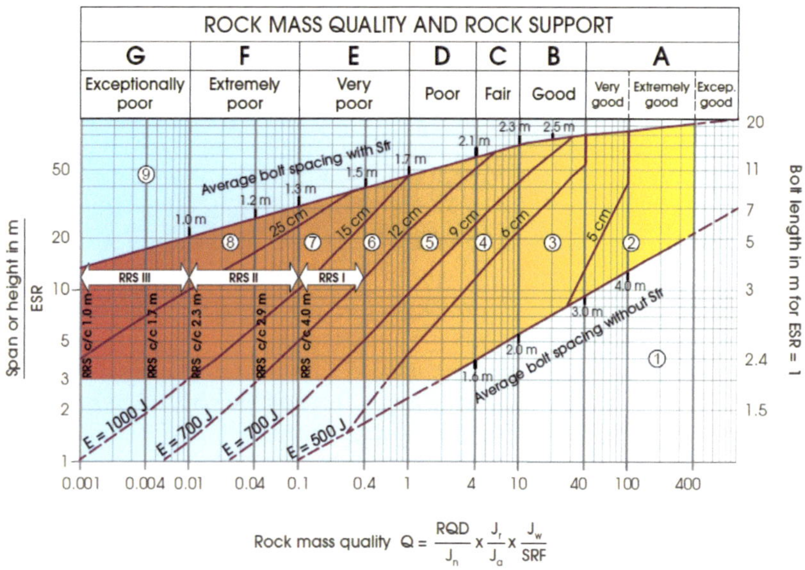

Figure 2.
Tunnel support chart based on Q-system [2].

for tunnels of diameter 5 m and 10 m have been given in **Table 1** for construction in various quality of rock mass with Q-values greater than 0.4. Construction of tunnels in rock mass less than Q-value of 0.4 would attract the large quantum of supports and hence the mining cost may not be feasible for the ores of an average value. For the ores containing the high mineral value (silver, gold, zinc etc.), hill mining using tunnels would be economical even in the rock mass of very poor quality which requires more stiff supports to stabilize the tunnel (s).

Category	Type of structure	ESR
A	Temporary mine openings, etc.	3–5
B	Vertical shafts*: i) circular sections ii) rectangular/square section * Depends on purpose and may be lower than given values.	2.5 2.0
C	Permanent mine openings, water tunnels for hydropower (exclude high-pressure penstocks), water supply tunnels, pilot tunnels, drifts and headings for large openings	1.6
D	Minor road and railway tunnels, surge chambers, access tunnels, sewage tunnels, etc.	1.3
E	Powerhouses, storage rooms, water treatment plants, major road and railway tunnels, civil defense chambers, portals, intersections, etc.	1.0
F	Underground nuclear power stations, railways stations, sports and public facilitates, factories, etc.	0.8
G	Very important caverns and underground openings with a long lifetime,≈ 100 years, or without access for maintenance.	0.5

Table 2.
ESR values assigned to various type of structures [2].

Tunnels or cross-cuts of 5 m diameter are very safe without any additional supports (except some spot bolting) for the suggested range of rock mass quality, i.e. $Q \geq 0.4$ (**Table 1**). Rate of mineral/ore production shall be low using tunnels/cross-cuts of 5 m diameter as compared to 10 m diameter tunnels because of deployment of smaller size machines in the former case. Similarly, mining with large tunnels of 10 m diameter shall also not require any additional supports, if excavation is carried out using controlled blasting technique i.e. causing minimum disturbance to the surrounding rock mass and the mined-out area is temporary and planned to be back-filled. Further, permanent openings of diameter 10 m created in a rock mass with quality index Q = 0.4–4 would need supports in form of 2.5 m long steel rock bolts at the spacing of 1.5–2.1 m with 6-9 cm steel fiber-reinforced shotcrete (SFRS). The same large openings in the rock mass with quality Q = 4–10 would need only rock 2.5 m long rock bolts at a spacing of 2 m as supports, whereas supports would not be required for the permanent 10 m diameter openings in a rock mass with $Q > 10$ (**Table 1**).

In mines, size of underground roadway i.e. roadway height and its dimension, as per the statute, is one point which needs due consideration while planning hill mining by tunneling method. In India, the dimensions of the pillars are regulated by Regulation 111 to 115 of Coal Mines Regulations (CMR), 2017 [4]. The regulation stipulates that the width of galleries shall not exceed 4.8 m and height of galleries shall not exceed 3 m. In this view, it is suggested to design the tunnels with width ≤ 4.8 m and height ≤ 3 m for coal deposits. Tunnels of larger dimensions may be designed for hill mining of non-coal minerals i.e. metallic minerals, which occur in narrow forms. However, smaller dimensions of excavation are always desirable from stability viewpoint because lager dimension gives rise to deformation and attracts requirement of stiffer supports. Thus, the smaller the opening better will be stability. If the value of the ore is high, the extraction by underground means, remain economical even after provision of stiffer support, mine planner can go for a larger dimension of underground openings.

For coal deposits having 30 x 30 m^2 pillar dimension, the strength of pillar lies in the range of 5.4 MPa to 7.4 MPa at various depths. The pillar width shall be according

For coal strength in a laboratory, σ_c = 5.4 MPa				
σ_s (MPa)	**W/h**	**σ_p (MPa)**	**H (m)**	**Factor of safety = (σ_s/ σ_p)**
9.68	3.2	6.08	100	1.59
11.85	4.32	7.61	150	1.56
13.72	5.28	9.17	200	1.50
15.90	6.4	10.55	250	1.51
18.07	7.52	11.91	300	1.52
20.25	8.64	13.27	350	1.53
22.12	9.6	14.70	400	1.50
24.30	10.72	16.05	450	1.51
26.16	11.68	17.45	500	1.50
For coal strength in the laboratory, σ_c = 7.4 MPa				
11.56	2.56	7.13	100	1.62
13.69	3.36	8.83	150	1.55
15.82	4.16	10.35	200	1.53
17.95	4.96	11.81	250	1.52
20.08	5.76	13.23	300	1.52
22.21	6.56	14.62	350	1.52
24.34	7.36	16.01	400	1.52
26.47	8.16	17.38	450	1.52
28.61	8.96	18.75	500	1.53

Note: All notations i.e. W, h, H, σs and σp has the same meaning as explained in Eq. 1 and 2 above.

Table 3.
Width of square shape coal pillars at various depths.

to the values given in **Table 3**. The gallery width and height are 4.8 m and 3 m respectively. Strength of pillar has been determined using the formula suggested by Bieniawski and Van [5] for square-shaped coal pillars (Eq.1).

$$\sigma_s = \sigma_c \left[0.64 + 0.36 \frac{W}{h} \right] \tag{1}$$

where,
σ_s = strength of pillar (MPa).
σ_c = unconfined compressive strength (MPa).
W = width of coal pillar (m).
h = height of pillar (m) or gallery.

Stress on the pillar as shown in **Figure 3**, is computed according to the following Eq.2:

$$\sigma_p = \frac{\gamma H.(W + D)^2}{(W)^2} \tag{2}$$

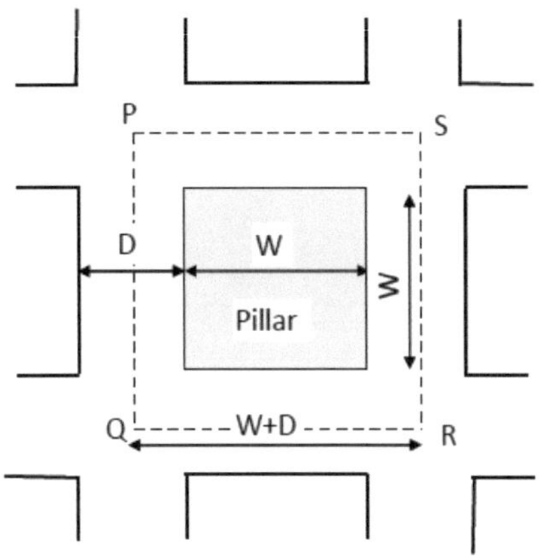

Figure 3.
Plan view of a coal pillar.

where,
σ_p = strength on the pillar (MPa).
γ = unit weight of rock mass above the pillar in MPa/m.
H = rock cover above the pillar.
W = width of the pillar (m).
D = width of the gallery (m).

D. Limitations and other issues

While dealing with hill mining by tunneling method, two important issues arise. Firstly, the 'shape of the underground openings' and secondly the mining methodology that includes both development and depillaring operation of mineral extraction. Due consideration should be given on both these points and its explanation has been given in the following paragraphs.

In general, tunnels are either D shaped (horseshoe shaped) or circular, having 'arched roof' whereas underground mine galleries have nearly flat roofs. Tunnels with an arched roof are more stable and hence safer compared with the underground galleries. This is scientifically established that *flat roofs* having corners have more chances of failure due to higher stress concentration at the corners. The mine planner can select rectangular openings, if depillaring and backfilling are the parts of the mining plan and method of mining because backfilling, after the depillaring operation is over, takes care of stress concentration at the corners which are the likely point of initiation or propagation of failure/crushing in the underground openings. Furthermore, in case of depillaring with backfilling, additional land area (on the surface) is not required for rock/waste storage generated from excavation underground. Therefore, dump creation and its management are eliminated, thereby economizing the overall cost of mineral production. Waste rocks which are huge in quantity destroy the landscape of the surface area completely and cause water and land pollution. Saving of the landscape of the mining area is priceless in terms of environmental benefits. In this way, the suggested tunneling method, which is an improved form of *underground mining method* would be both economically viable and environmentally friendly. Following are some limitations of 'tunneling methods' which shall be mentioned here.

- To make the 'tunneling method of mining' most feasible, the mined-out area shall be backfilled with rock waste, fly ash, sand or combination of more than one stowing material. Thus, 'depillaring with backfilling' is the best tunneling option, for safe exploitation of minerals by underground means.

- The method shall not be useful if entire rock cover is comprised of riverine material or soil having flowing tendency when destabilized.

- According to a rough estimate the cost of underground mining is more compared to the open -cast or surface mining [1] hence, cost analysis is essential for 'tunneling method of mining'. If found appropriate, this limitation can be easily overcome, particularly for high value and strategic minerals (gold, uranium ore, nickel ore, base metals i.e. Cu, lead, silver, zinc etc.).

When the tunneling method is practised, the value addition of the 'developed underground space' is suggestive. With such practices, the cost of production can be minimized and both revenues, as well as employment for locals, can be generated. Such value addition of the underground space/areas, if planned for the future, tunnels with an arched roof are advised. Research by the authors shows that the development roadways (areas developed during mine development, in particularly) has other civic uses too e.g. development as an underground storage space, place for a miscellaneous purpose, place of tourist extraction etc. [6].

3. Construction of approach road

Approach roads are vital at hill sites. The paucity of land and constrained space in hills are some well-known problems of hill mining. For open surface mines, larger length of approach roads is needed, If the tunneling method is the choice of mineral winning, road length is reduced and dust hazard is kept contained. **Figures 4** and **5** show various approach roads to tunnels excavated in hills. The approach roads, only means of hill transportation, shall remain functional round the year to keep geared the mining activities in the hills and also for transportation of man, material and machinery. Their construction and design should be rugged enough as per the load they have to handle because heavy machinery of big size will often use the roads.

Figure 4.
A view of the construction of approach road in the left for the tunnel shown on the right.

Figure 5.
View of approaches to the tunnel constructed in the hill.

3.1 Layout of tunnels and cross-cuts for mining

In a normal underground mine, the position of working face and its layout has close linkage as these facilitate the overall development of mine workings. Dimension and layout of tunnels and cross-cuts should be decided based on the thickness of the ore body. Thus, for the massive thickness of mineral bed, larger dimensions may be permitted. In case of metallic ore deposits which is found in narrow veins, lodes and pockets smaller dimensions suffice the purpose. For an ore-body having a thickness in the range of 5 m or more, development of mine working shall be through tunnels and cross-cuts of diameter 'D' diameter (**Figure 6**).

The tunnels are constructed within the orebody or in host rocks. Construction of underground openings in host rocks is a non-productive exercise and shall be kept limited. This methodology starts giving production right from the beginning when underground excavations are made in mineral instead of host or country rocks. The tunnels are interconnected with cross-cuts of the same or different cross-section perpendicular to the tunnel direction/alignment at a fixed interval. Thus, pillars

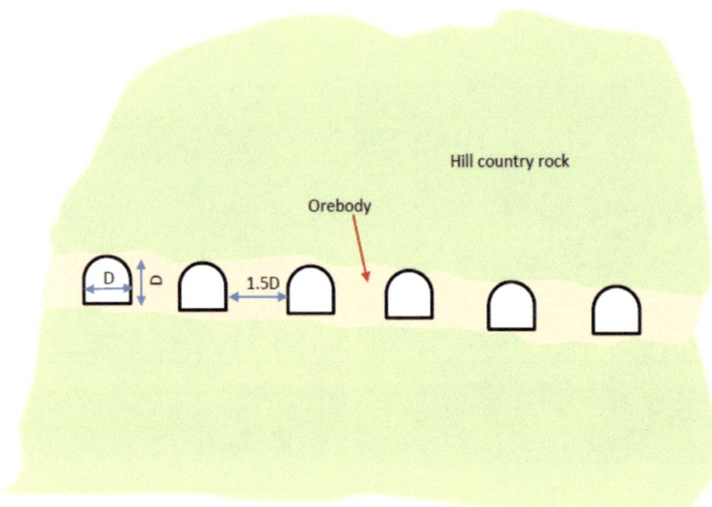

Figure 6.
Layout of tunnels for hill mining.

are formed which are extracted in the extraction phase of the mining operation. To get the mineral finally, these developed mine-workings are either fully or partially excavated.

The development of underground space, through tunneling and cross-cuts, for mining of minerals, has yet another dimension of multi-level mineralization in different horizons with host rocks in between, called 'parting'. All such levels, if not connected through tunnels from outside the hill, the dugout mineral or rocks can be poured down to the lower level and transportation underground can be done by gravity using shafts or chutes. Inter connectivity of levels at different points is required for the full development of the deposit which contains the mineral. If the mined-out area has to be backfilled then the pillar width should be reduced to a minimum with due consideration of roof stability. As such, the maximum quantity of ore should be dug out for mineral conservation purposes. At the decommissioning phase of 'eco-friendly hill mining by tunneling method', the underground space created during mining can create a value too and shall be maintained till the end. If done so, the underground space may be developed as a tourist place, underground market and other utilities like hotels/shops in future which has value.

Thus, two aspects of the discussed method are quite apparent, firstly, 'clean mining and green environment' and secondly, 'value addition' after mining is over. Revenue generation and socio-economic gains are therefore well connected to the method at once the rehabilitation of the created underground space is sought.

3.2 Productivity, risk and safety

In general, productivity is a measure of performance or output. It is a measure of how effectively the business targets of mining companies are being met [7]. In hill mining by underground approach, productivity is a ratio of input Vs output and implies how much mineral/ore is produced through various inputs. Obviously, hill mining, done by any method, has limited productivity and high risk. Its principal reasons are terrain condition. Another risks associated with the hill, particularly the Himalaya region, are the extreme weather conditions and environmental fragility, which should be attended scientifically.

Tunneling methods of mineral extraction has environmental and safety advantages over conventional mining. Safety in mines and mining industry (from accident angle) is prime and has to be maintained and accrued through constant efforts. At ground level, the safety can be enhanced or dealt effectively with knowledge of "Safety Management and Safety Engineering", as they are the modern and newly emerged tools to achieve the road to zero harm [8]. Undoubtedly, safety is a major concern for lofty hills and for this mining professionals are required to keep an eye on the stability of underground openings and slopes near the underground entry points i.e. portals. For this, scientific tools of numerical modeling, field measurement etc. pave the way as described in the following two paragraphs.

Deformation monitoring of excavated underground stretches: For safe underground working, the stability of the tunnel/cross-cuts is highly recommended to monitor so that counter measures can be taken in time to strengthen the rock mass around the excavated opening i.e. tunnel periphery wherever needed. For this, bi reflex targets are fixed at various cross-sections, especially near the cross-cuts because there are large spans of excavated space at such places, which need special care. After all, large excavations attract high induced stresses around the boundary of the openings. The targets are suggested to fix as given in **Figures 7** and **8**. One target at each TP1 (crown), TP2 & TP3 at the springing level and TP4 & TP5 at the upper bottom.

Frequency of such measurement depends on the trend of change in the radial deformation. If deformation shows an increasing trend, the frequency of

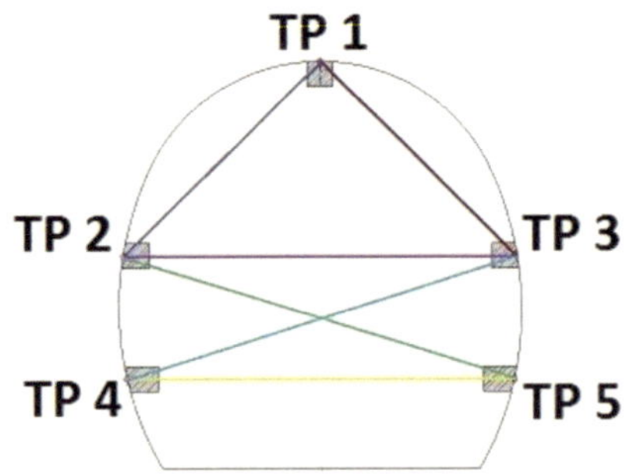

Figure 7.
Position of bi-reflex targets required to fix for monitoring radial tunnel deformation.

Figure 8.
Bi-reflex targets fixed to monitor radial tunnel deformation.

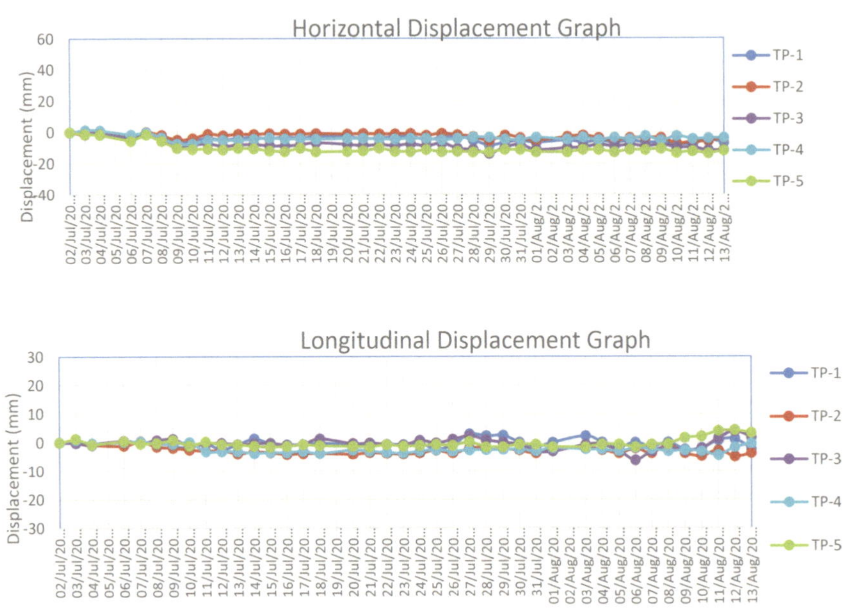

Figure 9.
Tunnel deformation recorded at all bi reflex targets of a tunnel section.

measurement needs to be increased. The readings obtained from such monitoring should be religiously analyzed so that action can be taken in advance before any untoward incidence (accident) takes place. **Figure 9** shows tunnel deformation in vertical, horizontal and longitudinal directions at TP1, TP2, TP3, TP4 and TP5 target points of a tunnel section. Alarming deformation level should be fixed at 80% of the allowable limit as per the design.

Slope monitoring near facade/tunnel portals: The purpose of a slope monitoring is to plan and maintain safe operating practises for the protection of personnel, equipment, and facilities. It provides warning of instability, so that action can be taken to minimize the impact of slope displacement and analyze the slope failure mechanism. The crucial geotechnical information provides help in designing the appropriate corrective measures [9]. Sufficient and suitable monitoring must be done to detect instability at an early, non-critical stage, to initiate the safety measures.

Opening of the mine i.e. portals of the tunnels is of prime importance as these serve as escape routes. Therefore, it is highly recommended to keep them intact and stable throughout the life of their existence and for a longer period, particularly when the mined-out areas have to be utilized for civic purposes in future. From this viewpoint, the portals are strongly supported with sufficient stretches (up to 15 m from the portals) and support density (1.25 times). To monitor the slope stability, bi-reflex targets are fixed at portal slopes as shown in **Figure 10**. Inclinometer and slope monitoring electronic gazettes such as 'terrestrial laser scanner (TSL)' and 'slope stability radar' (SSR) are some available equipment that may be used to assess the movement of hill slopes [10]. Depending on the requirement and feasibility on cost economics basis, they may be deployed. Piezometers can be installed to monitor the water head.

Precisely, the tunneling method for hills is an eco-friendly method that adopts similar basic principles as that of 'underground mining' in general and follows similar safety routes. To get a speedy and fast return on investment, the tunneling method of mining is best suited for the high-value minerals however, coal and other low-value minerals namely limestone, dolomite etc. can also be mined by this method taking into account the cost–benefit analysis. To make the method more cost-effective for mineral extraction, another dimension that may be added to this method is the 'eco-friendly transportation in hills' [11] and 'best practice mining' [1].

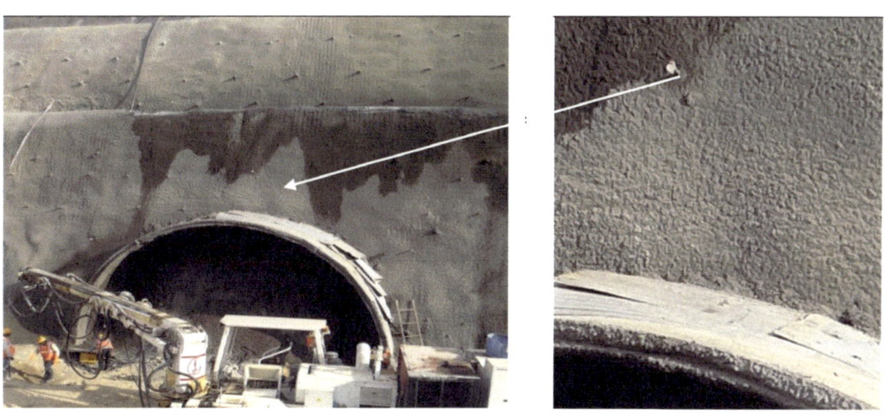

Figure 10.
Bireflex targets fixed at a tunnel portal to monitor the movement of the slope.

4. Case study: limestone mining in Meghalaya

Meghalaya, a high rain intensity state of India, in the North-Eastern part is mostly a hilly state (West, South and East Garo Hills; West and East Khasi Hills and Jaintia Hills) where sky seldom remains free of clouds. Meghalaya is rich in minerals and blessed with about 9% of the total limestone reserve of India [12]. The hills containing limestone minerals are being mined by open cast mining method in Meghalaya. Conventional 'Strip mining' is the method adopted in hill mining at Meghalaya.

Geologically, limestone in Meghalaya falls under the rocks formations namely Cretaceous-Tertiary sedimentary rock, which is further divided into three groups i.e., the Khasi group, the Jaintia Group and the Garo group. The Jaintia Group is further sub-divided into three formations, which include the *Longpar (lower), the Shella (middle)* and the *Kopili (upper)* formations.

The limestone deposited in Jaintia Hills, possesses limestone with alternating bands of sandstone. However, the limestone deposit in Cherrapunjee area of Meghalaya consists of limestone layers in the upper part of hills and dolomite in the lower portion. The limestone rocks found in Meghalaya belong to the Shella formations of the Jaintia Group of Cretaceous-Tertiary sedimentary rocks of Eocene geological age [3, 13].

As described above, in hill districts of Meghalaya, limestone is being extracted by open cast method of mining at both large scale and small scale levels. Jaintia Hills is being extracted in large scale for cement, whereas East Khasi hills are being extracted in small and large scale for manufacturing of quick lime, edible lime and cement. Most of the mines are owned by small private entrepreneurs. Some of the landowners are organized, and some make use of crude methods and adopt unscientific practices of mining on an individual basis to extract limestone. The captive mines of the cement industries are efficient, being mechanized, make use of heavy machinery for excavation. On the other hand, extraction by individual landowners is manual or semi-mechanical only and thus slow.

It is noticed and revealed periodically, that the local environment around the mines or in mining regions has been affected by creating hullabaloo. Engaged mining companies of the concerned area of the state faces its consequences both financially and socially (**Figure 11**). In general, extraction of limestone involves mechanical removal of overburden (using bulldozers), drilling of the blast holes, blasting of rocks (shattering), sizing, loading and then finally transportation of limestone to the consumer or industry cement plants.

In many hilly areas of Meghalaya, quarrying, for limestone, building stone/ material such as slate, granite, clay etc., is a typical sight. Most of them are unscientific and cause disastrous and irreversible changes to natural habitats. At the

Figure 11.
View of various opencast mines in Meghalaya.

developmental phase, to reach the deposit, the mineral winning process shall be through driving of tunnel and approaching underground instead of the surface.

With adequate planning, 'the tunneling method' can be implemented into practices at Meghalaya as its tremendous feasibility exist. In this way, agricultural land and the landscape, nearby rivers and other water bodies, are not polluted. Air, water and land environment of the area can be protected with underground hill mining approach (tunneling).

5. Conclusions

Our experience of working in Indian mines and the analysis described in this technical communication concludes that 'the conventional hill mining can be turned into eco-friendly mining with small efforts, according to the hill topography, when the tunneling method is selected for implementation into practice. Many hill areas, including The Himalayas, will be the direct beneficiary and by doing so land degradation could be reduced to a minimum. The ill-effects of surface mining e.g. possibility of deforestation/denudation of forest, creation of scars on hill slopes (defacing) due to dumping on slopes, destabilization of natural hill slopes, landslides (destabilization) of hills, water pollution and disturbance to natural drainage pattern of the hills are either eliminated or curbed substantially. In this way, mining and environment can go hand-in-hand and the greenery of a hill and the surrounded environment is preserved. The less disturbed land surface, on one hand, protect the serene hill environment and on other hand allows the production of mineral deposits in hills irrespective of the scale of mining.

Author details

Rama Dhar Dwivedi[1] and Abhay Kumar Soni[2]*

1 CSIR-Central Institute of Mining and Fuel Research (CIMFR), Roorkee, (Uttarakhand), India

2 CSIR-Central Institute of Mining and Fuel Research (CIMFR), Nagpur, (Maharashtra), India

*Address all correspondence to: abhayksoni@gmail.com

References

[1] Soni A.K. (2017), Mining in the Himalayas - An Integrated Strategy, Book Published by CRC Press / Taylor & Francis, p. 225 (ISBN - 9781498762342).

[2] NGI (2015). Using the Q-system: Rock mass classification and support design, Norwegian Geotechnical Institute (NGI) Publication, 54p.

[3] DMR (2016), Directorate of Mineral Resources (DMR) Profile, Government of Meghalaya (http://megdmg.gov.in/), Shillong, Meghalaya.

[4] GOI (2017), Coal Mines Regulations, 2017, Government of India (GOI), Amended and modified from Coal Mines Regulation (CMR), 1957(from the web; last accessed 0n 17/09/2020).

[5] Bieniawski ZT, Van HWL (1975). The significance of in situ tests on large rock specimens. Int. J. Rock Mech. Min. Sci. Geomech. Abstr. 12(4): 101-113.

[6] Soni A.K., (1995), Environmental Study of a Limestone Mining in the Himalayas, Journal of Mining Research, Vol. 3, Nos. 3 &4, January–March, pp.1-8.

[7] Jairo Ndhlovuand Peter R K Chileshe (2020), Global Mine Productivity Issues: A Review, Int. J. of Engineering Research & Technology (IJERT), Vol. 9 Issue 05, May (ISSN: 2278-0181; http://www.ijert.org), pp. 319-329.

[8] Soni A.K. and Suman Kiran (2012), The Road to Zero Harm: Safety Management and Safety Engineering in Context to Underground Mines, Workshop on Safety Management in Mines. Nagpur, pp. 01-10.

[9] Call R.D., J.P. Savely (1990). Open-pit rock mechanics. Surface Mining, 2nd Edition. Society for Mining, Metallurgy and Exploration, Inc., pp. 860-882. B.A. Kennedy ed.

[10] Prakash Amar, Kumar A., Singh K. B. (2016). Highwall mining: A critical appraisal, MineTech, Vol. 36 No. 3, pp. 17-30.

[11] Soni A.K. (1999), Environmentally - Friendly Transportation of Limestone for Cement Production, International Journal of Bulk Solid Handling - Germany, ISSN: 01739980 Vol.19. N0.3, July–September, pp. 329-336.

[12] Lamare R. E., Singh O. P. (2016). Limestone mining and its environmental implications in Meghalaya, India. ENVIS Bulletin Himalayan Ecology, 24:87-100.

[13] Sarma S. (2003). Meghalaya: The Land and Forest - A Remote Sensing Based Study, Geophil Publishing House, Guwahati, India, 5-16.

Chapter 4

Reclamation of Soils Degraded by Surface Coal Mining

Luiz Fernando Spinelli Pinto, Lizete Stumpf, Pablo Miguel, Leonir Aldrighi Dutra Junior, Jeferson Diego Leidemer, Lucas da Silva Barbosa and Mauricio Silva e Oliveira

Abstract

The largest Brazilian coal mine, called Candiota mine, is located in South Brazil, with an estimated reserve about 1.2 billion tons. Since late 2003, an experiment located at a reclaimed site in a coal mining area was conducted, in which a research group from the Federal University of Pelotas has been conducting a long-term experiment on soil quality with different plants species, such as *Hemarthria altissima*, *Paspalum notatum* cv. Pensacola, *Cynodon dactylon* cv. Tifton, and *Urochloa brizantha*. After 8.6 years of revegetation, soil samples at 0.20 depth were collected in minesoil and natural soil to determine physical attributes, and the organic carbon content. After 10.9 years of revegetation, soil samples at 0.10 m depth were collected to determine the biological attributes. According to the research results, it can be seen that the recovery of minesoil was more effective after 8.6 years of revegetation only in the physical condition up to 0.10 m depth. However, all soil physical attributes and organic matter content are still below the levels observed in the natural soil. The biological attributes after 10.9 years of revegetation have not yet been sufficient to restore a mites and springtails population close to the natural soil.

Keywords: minesoil, revegetation, physical attributes, organic matter content, edaphic mesofauna

1. Why this study is important?

"Soil" is borne as a result of lengthy natural processes over thousands of years; hence, it is a valuable nonrenewable commodity. It is a basic environment needed for vegetation growth on land, be it a mined land or other. In case of soils degraded by surface coal mining, one should not bear in mind it would be a simple task to bring back degraded/mined soil to its near original configuration so that it would become naturally capable to sustainably support vegetation. With this aim, we carried out our study and here lies the "time period to bring back the degraded minesoil to close to natural soil condition," which is an extremely important requirement for surface coal mining successful closure. This research study has put stress on the long-time scientific evaluation of coal mine soil degraded by the excavation operation, i.e., mining (for more than 16 years). Though maintaining such experiment requires lot of efforts and resources, we think it is a necessary tool to analyze the question we have just put forward.

Our study was done in a randomized block design field experiment, sampling the same soil over the time, and comparing the soil properties with the natural soil, what makes the data obtained more scientifically reliable and meaningful. We, as authors, have tried to make this idea more clear in our writing; it is well known and obvious that the soils properties once covered with vegetation will tend to improve over time. Nevertheless, this does not necessarily happen, and sometimes, many sites show signs of degradation and even erosions problems after many years of reclamation, needing re-intervention.

Therefore, the main difference between our study and other similar studies is that of experimental control. Most studies deal with sampling of mining sites, with different ages, but without experimental control. It is also important to do research on soil reclamation techniques and procedures focusing on improving minesoil quality, ensuring the return of a productive soil according to the planned use.

2. Soils formed in surface coal mining

Coal remains a major fuel in global energy systems, accounting for almost 40% of electricity generation, and over the next 5 years, the global coal demand is forecasted to remain stable, supported by the resilient Chinese market, which accounts for half of the global consumption [1]. World coal reserves have a volume of approximately 860 billion tons, with deposits distributed in 75 countries. Of the existing reserves, 75% are concentrated in five countries: the United States, Russia, China, Australia, and India.

Brazil has one of the largest reserves of mineral coal in Latin America [2], and in recent years, it has been regaining its space in the energy market due to the need to supply the scarcity of electricity generated by water resources (due to seasonal lowering of water in the reservoirs). In Southern Brazil, the largest deposit in the country called Candiota Deposit is located, in which reserves of 1.2 billion tons are capable of being surface mined, at depths of up to 50 m [3].

The sequence of surface coal mining involves the previous removal of the original soil horizons, to then remove overburden rocks (**Figure 1a**,**b**, respectively). After coal seams extraction, the topographic reconstruction occurs, in which there is the return of the overburden rocks to fill the previous stripped area, and finally, the surface is leveled and topsoil is deposited to finish topographic recomposition (**Figure 1c**,**d**, respectively), creating an anthropogenic soil (**Figure 1e**).

Anthropogenic soils are soils that have been influenced, modified, or created by human activity. They are found worldwide in urban and other human-impacted landscapes. Four distinct types of anthropogenic soils can be distinguished based on geographical setting and historical context: (i) agricultural, (ii) archeological, (iii) mine-related, and (iv) urban [5]. According to the World Reference Base (WRB) [6], anthropogenic soils found in agricultural and archeological settings are classified as Anthrosols, whereas those in mine-related and urban settings are classified as Technosols. Anthrosols are formed by the transformation of natural soil by human additions of organic or inorganic materials over long periods of time, while Technosols are formed in parent materials created and deposited by human activities (e.g., mine spoils, urban fill). The most extensive mine-related anthropogenic soils are primarily associated with modern landscapes created by the surface mining of coal, and are classified as Spolic Technosols, according to the WRB, based on the fact that they contain technogenic artifacts in the form of mine spoil [5].

Before 1970s soil survey reports in the USA identified mined lands on maps and referred to them as mine dumps, mine spoils, or strip mines and mine-land reclamation its grouped surface materials on mined lands into various categories to assist

Figure 1.
Coal mining process (a-b) and topographic restoration (c-e) in southern Brazil [4].

with treatment for revegetation [7]. In the 1970s, after the passage of the Surface Mining Control and Reclamation Act (SMCRA) of 1977 and the resultant state permanent regulatory programs, coal mined lands were mandated to be returned as close as possible to the approximate original landscape, and since successful revegetation was rigorously required, natural topsoil, or a topsoil substitute (in case of the pre-1970s mining), was placed at the final reclamation surface [8]. Modern mining regulations also started to require the isolation of acid-producing (FeS_2) materials below the final surface. Since then, these soils, resulting from the reclamation process, have been called in the USA as minesoils [7, 9, 10], or less frequently as mine soils [8].

Minesoils, as the result of the mining and reclamation process, compared to the contiguous native soils, are much younger soils with properties more determined by human-controlled influences rather than by natural processes [9]. Their profile morphology can roughly be described as mainly composed of two layers, a surface layer made by the topsoil (the native soil A horizon) abruptly lying over a overburden layer. After few years of revegetation and exposure to climatic conditions, even in topsoil substitute layers, these young A horizons start to be loosened by root growth and organic matter accumulation and decomposition, developing color darkening and some soil structure. Also, the surface mining may accelerate the soil-forming processes by breaking up the consolidated rocks of the overburden layers allowing air, water, and plant roots to penetrate this layer [6]. Therefore, in strict pedological description of horizons, usually A-C horizon sequences in very young soils (<10 years old) and A-AC-C sequences in relatively older soils (>10 years old) are found. In some older profiles (>30 years old), the beginning of formation of B horizons (Cambic) has been reported [9].

The topsoil addition surely improves the minesoil quality, but heavy machinery traffic and inadequate soil distribution can hinder the vegetation development, the main starting point for the minesoils recovery [11]. As the consequence of excessive traffic from large machines during topographic recomposition, persistent topsoil compaction (**Figure 2a,b**) has been reported as a major impact on the physical quality of minesoils in India [12], in China [13], in UK [14], in South Africa [15], in Germany [16], in the USA [17], and in Brazil [18].

The development and evolution of the reclaimed minesoil provides a unique opportunity to expand the existing knowledge about the formation and stabilization of aggregates, accumulation and distribution of organic matter and microbial biomass, since, due to the magnitude of the disturbance of the ecosystem, it creates a sort of "zero time" scenario [19]. The success of the minesoil recovery does not only depends on the mining methods, the height and slope of the overburden piles, the nature of the mined soils, the geoclimatic conditions, but also depends on the plant species selected for their revegetation [20]. In this sense, a great number of plant species have been researched as an alternative to recover the quality of coal minesoils in different places in the world, some of which are cited below.

2.1 Reclamation of minesoils and revegetation in the USA

Soil and plant data among a chronosequence of 19 post-mine reclaimed sites (over a 40-year reclamation gradient), and an intact native reference site were evaluated. It was noticed that root biomass in the upper horizons (at 30 cm depth) was greater on the reference site compared with the reclaimed sites as well as the

a) b)

Figure 2.
Compaction of topsoil immediately after topographic restoration of the minesoil (a) and after 8.6 years of revegetation in southern Brazil (b) [4].

organic matter content, ranging from 3·5 to 5·4% on the reclaimed sites (not different across the reclamation chronosequence) and from 5·1 to 6·8% on the reference site [21]. On the other hand, in the Midwestern USA, there was the development of horizons in minesoils in a relatively short period of time (10–15 years), in which the 0.00–0.03 m layer consisted of non-decomposed or partially decomposed organic matter, while the 0.03–0.10 m layer was darker, with visible addition of organic carbon, and the 0.10–0.25 m layer was the least colored with interspersed roots [22]. When opting for the natural revegetation of mined areas, it was observed that minesoils up to 2 years of age have a predominance of annual and perennial grasses, while minesoils with 16–20 years usually have some tree species, and minesoils with 38–42 years old have a mix of native trees and understory species [23].

2.2 Reclamation of minesoils and revegetation in China

Vegetation succession and soil characteristics under five different restoration models of refuse dumps including different-aged revegetated sites were evaluated. It was observed that the biomass of the naturally species increased from 0.15 kg m^{-2} in the 8-year-old vegetation to 0.64 kg m^{-2} in the 18-year-old vegetation. Furthermore, the soil bulk density decreased from 1.56 Mg m^{-3} in 8-year-old vegetation on the abandoned land to 1.24 Mg m^{-3} for 18-year-old vegetation [24]. In another study, the minesoil showed improvements in its edaphic quality after 5 years of revegetation, which promoted an increase in the content of organic matter and a reduction in runoff and soil erosion [25]. On the other hand, the positive effects of revegetation on microbial activity were observed over 18 years of minesoil's formation [26].

2.3 Reclamation of minesoils and revegetation in other countries

In India, carbon dynamics in one unreclaimed site (0 years) and four chronosequences revegetated coal mine sites (3, 7, 10, and 15 years) were compared with an undisturbed forest as a reference site. It was verified that soil organic carbon stock significantly increased from 0.75 Mg C ha^{-1} in 3 years to 7.60 Mg C ha^{-1} after 15 years of tree species revegetation in the top 15 cm of soils [27].

In Spain, the effectiveness of using native colonizer shrubs as nurse plants to reintroduce the two main tree species present before the mining operations was evaluated. It was found that the seedlings mortality under shrubs increased during the second year after plantation, probably because of the lower precipitations during the second growing season that reduced the water holding capacity of then minesoil (1–3.5 g cm^{-2}) when compared with the nearby natural forest soil (19.8 g cm^{-2}) [28].

In southeastern Nigeria, minesoils under 30 years of natural revegetation still lacked an O horizon and high values of soil density in relation to natural soil [29].

In Germany, it was observed that minesoils after 4 years of revegetation still showed very variable physical properties, and that the choice of perennial species with deeper rooting was recommended to accelerate the formation of the new soil structure [16].

2.4 Reclamation of coal minesoils and revegetation in southern Brazil

Minesoils that use little topsoil thickness give rise to contamination with the fragments of overburden rocks, frequently showing high soil bulk density, lower macroporosity, and high mechanical resistance to penetration, in addition to spots with very low pH values (<3.0) [30].

Attributes of minesoil under the cultivation of *Hemarthria altissima*, *Paspalum notatum* cv. Pensacola, *Cynodon dactylon* cv. Tifton, and *Urochloa brizantha* were evaluated in a randomized block design experiment at 5 [31], 41 [32], 72 [33, 34], 78 [35], and 103 months [36]. The results are reported below:

a. At 5 months of revegetation, there were no differences in the attributes of the minesoil under the different species. However, the highest concentration of aggregates in the 0.00–0.10 m layer occurred in the 1.00–0.25 mm class (32.67%), while in the 0.10–0.20 m layer, the highest concentration occurred in the 4.76–2.00 mm class (26.68%). The average carbon content in the 0.00–0.10 m layer was 5.34 g kg^{-1} and in the 0.10–0.20 m layer, it was 5.18 g kg^{-1}.

b. At 41 months of revegetation, there were also no differences in soil attributes under the different species. However, the highest concentration of aggregates occurred in the 1.00–0.25 mm class, both in the 0.00–0.10 m layer (40.13%) and in the 0.10–0.20 m layer (35. 73%). The average organic carbon content in the 0.00–0.10 m layer was 7.38 g kg^{-1} and in the 0.10–0.20 m layer, it was 6.20 g kg^{-1}.

c. At 72 months of revegetation, in the 0.00–0.05 m layer, the lowest value of the pre-consolidation pressure was provided by *Hemarthria altissima* (71 kPa) while the other plant species showed higher values provided: *Paspalum notatum* cv. Pensacola (120 KPa), *Cynodon dactylon* cv. Tifton (120 kPa), and *Urochloa brizantha* (118 kPa).

d. Also at 72 months of revegetation, in the 0.00–0.03 m layer, it was observed that *Hemarthria altissima* and *Urochloa brizantha* provided the highest carbon stocks in the light free fraction (1.22 Mg ha^{-1} and 1.27 Mg ha^{-1}, respectively) compared to *Paspalum notatum* (0.86 Mg ha^{-1}) and *Cynodon dactylon* (0.83 Mg ha^{-1}). In relation to the carbon stock of the light occluded fraction, *Hemarthria altissima* and *Cynodon dactylon* presented higher stocks (1.09 Mg ha^{-1} and 1.02 Mg ha^{-1}, respectively) compared to *Paspalum notatum* (0.61 Mg ha^{-1}).

e. At 78 months of revegetation, it was found that concentration of macroaggregates was higher in the 0.10–0.20 m layer (87.56%) compared with the 0.00–0.10 m layer (81.15%). Average organic carbon content in the 0.00–0.10 m layer was 8.46 g kg^{-1} and in the 0.10–0.20 m layer, it was 6.39 g kg^{-1}.

f. After 103 months of revegetation, root's perennial grasses concentration and minesoil physical attributes were measured. It was verified that the root mass concentration ranged from 66 to 81% in the 0.00–0.10 m layer decreasing to 13–28% in the 0.10–0.20 m layer, due to inadequate physical conditions below the 0.00–0.10 m layer, indicated by macroporosity values below 0.10 m^3 m^{-3}, bulk density greater than 1.40 Mg m^{-3}, and the highest percentage of macroaggregates with large, cohesive, and sharp-edged aggregates features. In relation to this, a different soil-aggregation hierarchy path in clay minesoils with highly compacted topsoil was proposed, in which, prior to revegetation, compacted aggregates arising from the compression of the soil mass made by the intense movement of heavy machinery were produced during topographical recomposition. Thus, in the first year after revegetation, the 0.00–0.10 m soil layer presented smaller aggregates arising from the breakdown of the large cohesive aggregates than the 0.10–0.20 m layer. From this point on, aggregation would begin to develop with the action of decomposed roots and microorganisms favoring the conglomeration of particles, with sequential reformation and stabilization of aggregates, following the traditional soil-aggregation hierarchy path. As the root system progressively reaches and develops in the 0.10–0.20 m layer, the same process mentioned above is expected to occur. It is important to mention that all hierarchical levels mentioned above can occur simultaneously within the same layer of the minesoil.

3. Physical and biological attributes of minesoil revegetated with perennial grasses compared with the natural soil in southern Brazil: a case study at the Candiota coal mine

In late 2003, a field experiment located at a reclaimed site in the Candiota coal mine (31°33′56″ S and 53°43′30″ W) was implemented, under concession of the Riograndense Mining Company, and the research group from the Pelotas Federal University has been conducting a long-term experiment on the soil quality with different plants species.

The topsoil used to cover the coal overburden was composed mainly by the B horizon of the natural soil (prior to mining), a Rhodic Lixisol [6], with high clay content (466 g kg^{-1} clay), dark red color (2.5 YR 3/6), and lower organic matter content (12 g kg^{-1}) compared to the A horizon (21 g kg^{-1}). The experiment was installed in November/December 2003 in a randomized block design with four replicates (each plot with 4 × 5 m = 20 m^2). Grasses used as treatments consisted of perennial summer grasses (**Figure 3**): *Hemarthria altissima* (15 cuttings m^{-2}), *Paspalum notatum* cv. Pensacola (50 kg of seed ha^{-1}), *Cynodon dactylon* cv. Tifton (15 cuttings m^{-2}), and *Urochloa brizantha* (10 kg of seed ha^{-1}). Prior to the implantation of the cover crops, the soil was chiseled with a bulldozer up to 0.15 m depth, and also received dolomitic limestone equivalent to 10.4 Mg ha^{-1} effective calcium carbonate rating and 900 kg ha^{-1} of NPK fertilizer, 5-20-20 (45 kg N, 180 kg P_2O_5, and 180 kg K_2O). Annually, all plots received 250 kg ha^{-1} of NPK fertilizer, 5-30-15 (12.5 kg N, 75 kg P_2O_5, and 37.5 kg K_2O) and 250 kg ha^{-1} of ammonium sulfate.

In July 2012 (8.6 years of revegetation), soil samples in the 0.00–0.10 m and 0.10–0.20 m layers were collected in minesoil and natural soil to determine the granulometry [37], tensile strength [38, 39], distribution of water stable aggregates in size classes [40, 41], bulk density and soil porosity, and the organic carbon content [42]. In October 2014 (10.9 years of revegetation), the soil samples in the 0.00–0.10 m layer were collected to determine the microbial biomass carbon [43], metabolic quotient [44], and organisms of the edaphic mesofauna, represented by mites and springtails [45]. All soil attributes differences were compared to the natural soil under native vegetation (reference soil).

The predominant natural soil of the mining area is a Rhodic Lixisol with 477.79 g kg^{-1} sand, 271.81 g kg^{-1} silt, and 250.40 g kg^{-1} clay in the 0.00–0.10 m layer, and 444.91 g kg^{-1} sand, 256.09 g kg^{-1} of silt, and 299.00 g kg^{-1} of clay in the 0.10–0.20 m layer [4]. Due to the soil construction processes, both the 0.00–0.10 and 0.10–0.20 m layers of the minesoil present, respectively, 80.91 and 59.87% higher clay content (453 and 478 g kg^{-1}, respectively) than the non-anthropized natural soil. Differences in clay content can make attribute comparisons between minesoils and natural soils questionable, as higher clay contents contribute to greater aggregation through the reorientation of clay particles, binding with root exudates and wetting and drying cycles. By contrast, measuring soil attributes prior to coal mining allows one to understand the intensity of the impact of mining

Figure 3.
Hemarthria altissima (a), Paspalum notatum cv. Pensacola (b), Cynodon dactylon cv. Tifton (c), and Urochloa brizantha (d) implanted in minesoil in southern Brazil [4].

on the environment. Consequently, the differences between the attributes of the natural and the minesoil are important in estimating the recovery period required for the new soil profile to perform functions in the environment in which it is inserted.

In this sense, after 8.6 years of revegetation, it is possible to observe that the minesoil under *Urochloa brizantha* and *Paspalum notatum* presented in the 0.00–0.10 m layer, respectively, 1.8 and 5.7% lower percentages of macroaggregates, while the constructed soil under *Hemarthria altissima* and *Cynodon dactylon* presented, respectively, 2.4 and 3.5% higher percentages of macroaggregates in relation to the natural soil (89.15%). In the 0.10–0.20 m layer, the treatments presented 16.4–19.2% higher percentage of macroaggregates in relation to the reference soil (80.65%) (**Figure 4a**). The largest proportion of macroaggregates presented by minesoil below the 10 cm layer, relative to natural soil, does not refer to a natural aggregation process promoted by biological forces (roots and exudates of microorganisms), but formed by the compression generated by intensive machines traffic during the topographic recomposition of the mined area [36].

Regarding the percentage of microaggregates, it was observed that in the 0.00–0.10 m layer, *Urochloa brizantha* and *Paspalum notatum* promoted, respectively, 46.9 and 14.9% higher percentage, while *Hemarthria altissima* and *Cynodon dactylon* promoted, respectively, 19.5 and 18.5% lower percentage than the reference soil (10.85%). In the 0.10–0.20 m layer, the treatments presented 68.6–80.2% lower microaggregation than the natural soil under native vegetation (19.35%) (**Figure 4b**). In a minesoil in the USA [39], macroaggregation was 50% smaller and microaggregation was 10% smaller in less than 1 year old soil (64% sand, 22% silt, and 19% clay) when compared to natural soil (55% sand, 29% silt, and 16% clay). However, after 16–20 years of revegetation, there was similarity between the distribution of minesoil aggregates (56% sand, 31% silt, and 13% clay) in relation to soils not disturbed by coal mining (59% sand, 28% silt, and 13% clay).

Figure 5 shows that in the minesoil under the perennial grasses, the aggregates presented 24.9–66% higher tensile strength compared to the natural soil (55.98 kPa) in the 0.00–0.10 m layer, while in the 0.10–0.20 m, the tensile strength values of the treatments were 163.9–221% higher than the reference soil (66.28 kPa). Similar results in a coal minesoil after 2.8 years of revegetation was observed, with higher tensile strength values in the 0.00–0.05 m (70.32–88.81 kPa) and 0.10–0.15 m (70.90–125.92 kPa) layers of grass covered in comparison to the natural soil under

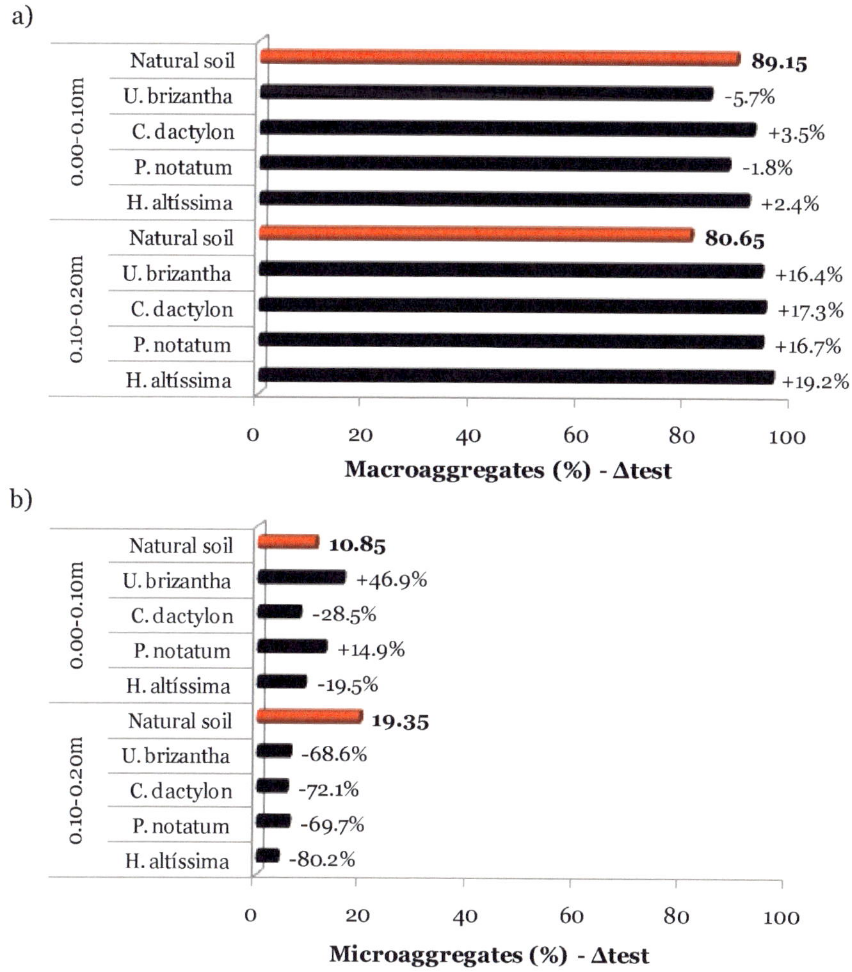

Figure 4.
Differences (Δtest) between percentage macroaggregates (a) and microaggregates (b) of minesoil after 8.6 years of revegetation (under perennial grasses) relative to natural soil (under native vegetation).

native vegetation (0.00–0.05 m: 55.98 kPa, and 0.10–0.15 m: 66.28 kPa) [46]. The higher tensile strength of aggregates is due to the effect of machine traffic during the topographic recomposition of the area, which resulted in cohesive, hard, and poor porous aggregates [38].

After 8.6 years of revegetation, it was also observed that the minesoil under different perennial grasses presented soil bulk density up to 21.1% higher in the 0.00–0.10 m layer, while in the 0.10–0.20 m layer, the difference was 15.7–34.05% in relation to the natural soil under native revegetation (presented 1.20 Mg m^{-3} and 1.18 Mg m^{-3}, respectively) (**Figure 6**). This result is due to topsoil compaction during the topographic recomposition of the mined area, commonly cited in the literature [47]. On the other hand, other studies indicate that the bulk density decreases over time, as observed in a minesoil in the USA, which is presented at 5, 10, and 16 years of revegetation values of 1.82 Mg m^{-3} (69% sand, 21% silt, and 10% clay), 1.70 Mg m^{-3} (50% sand, 28% silt, and 22% clay), and 1.48 Mg m^{-3} after 16 years (44% sand, 32% silt, and 24% clay). However, even after 16 years of revegetation, the bulk density was higher than the natural soil under grass (1.26 Mg m^{-3}) [48].

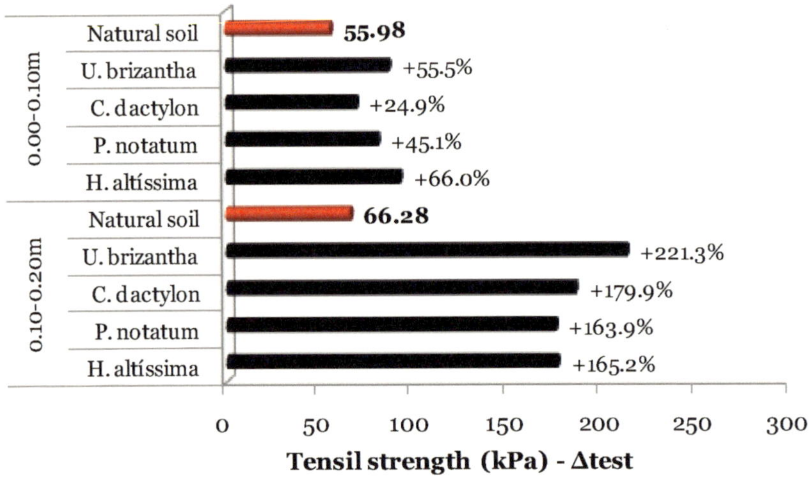

Figure 5.
Differences (Δtest) between tensile strength aggregates of minesoil after 8.6 years of revegetation (under perennial grasses) relative to the natural soil (under native vegetation).

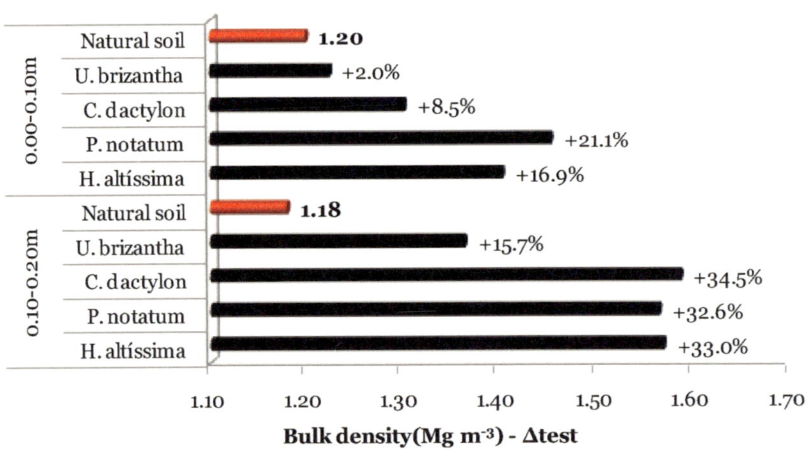

Figure 6.
Differences (Δtest) between the bulk density of minesoil after 8.6 years of revegetation (under perennial grasses) relative to the natural soil (under native vegetation).

When evaluating pore distribution, it was observed that in the 0.00–0.10 m layer, the minesoil under *Urochloa brizantha* and *Cynodon dactylon* presented, respectively, 26.4 and 25.9% higher macroporosity than the natural soil under native vegetation, while the other species presented lower values, highlighting the potential of the root system of these species, which presented in the layer of 0.00–0.10 m 92 and 93% of their roots with a diameter smaller than 0.49 mm [18]. However, below the 0.10–0.20 m layer, it was observed that the treatments presented 4.9–70.5% lower macroporosity than the reference soil (**Figure 7**), which was the consequence of the higher degree of compaction of minesoil.

The results presented show the difficulty in revegetating mined areas and, consequently, in allowing the natural incorporation of organic waste in the minesoils [49], which directly influences the regeneration of these areas. **Figure 8** shows that the organic carbon content of the minesoil was 48.3–58.2% lower in the 0.00–0.10 m layer compared to the natural soil (20.04 g kg^{-1}), while in the 0.10–0.20 m layer, the values were 18.6–53.1% lower than the natural soil (10.26 g kg^{-1}).

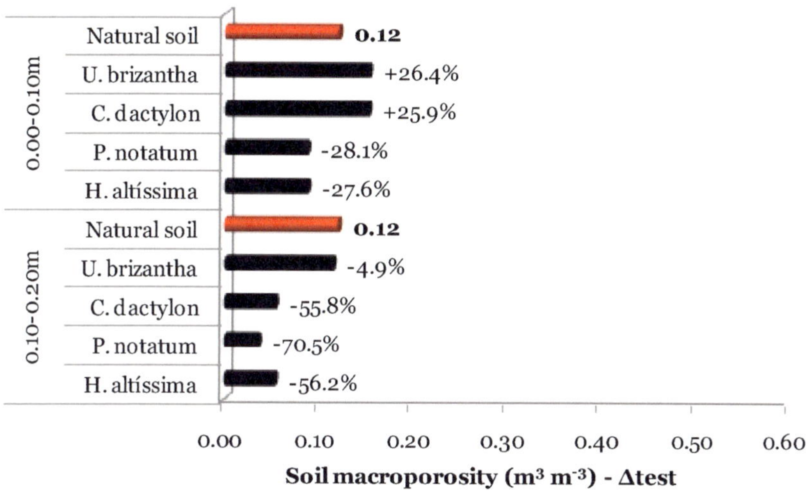

Figure 7.
Differences (Δtest) between the macroporosity of minesoil after 8.6 years of revegetation (under perennial grasses) relative to natural soil (under native vegetation).

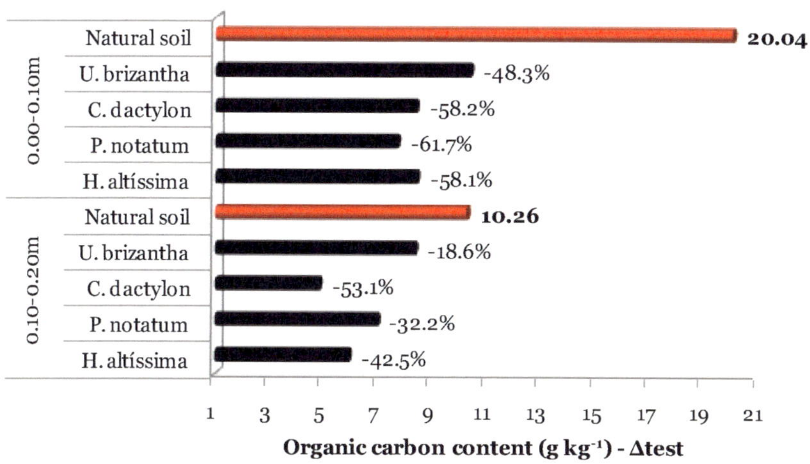

Figure 8.
Differences (Δtest) between the organic carbon content of minesoil after 8.6 years of revegetation (under perennial grasses) relative to the natural soil (under native vegetation).

However, in a minesoil in the USA, higher levels of organic carbon were observed in minesoils after 14 years (19.7 Mg ha^{-1}) and 26 years (13.4 Mg ha^{-1}) of revegetation than the natural soil (9.92 Mg ha^{-1}) [48].

The higher carbon content in natural soil is linked to the presence of microorganisms in the soil. In this sense, after 10.9 years of revegetation, it was observed that the natural soil presented 373 g $kg^{-1}$of microbial biomass carbon in the 0.00–0.10 m layer. The minesoil under the different grasses presented up to 42.69% lower values, except the soil under *Hemarthria altissima*, which presented values similar to the natural soil (**Figure 9a**). This result highlights the importance of adding carbon sources in recovering areas, aiming at the improvement of biochemical conditions, which may favor the return of soil biological balance.

On the other hand, **Figure 9b** shows that the metabolic quotient (qCO_2) values in the 0.00–0.10 m layer of the minesoil were 23.4 and 103.1% higher than the natural soil (0.64). A high qCO_2 indicates that the microbial population is experiencing

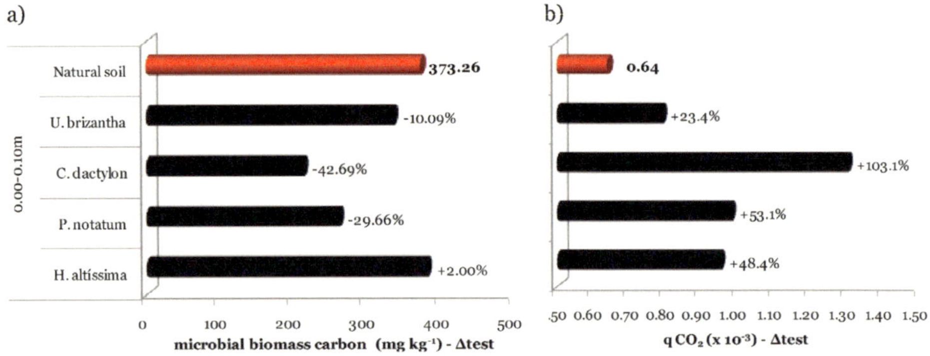

Figure 9.
Differences (Δtest) between microbial biomass carbon (a) and metabolic quotient (b) of minesoil after 10.9 years of revegetation (under perennial grasses) relative to natural soil (under native vegetation).

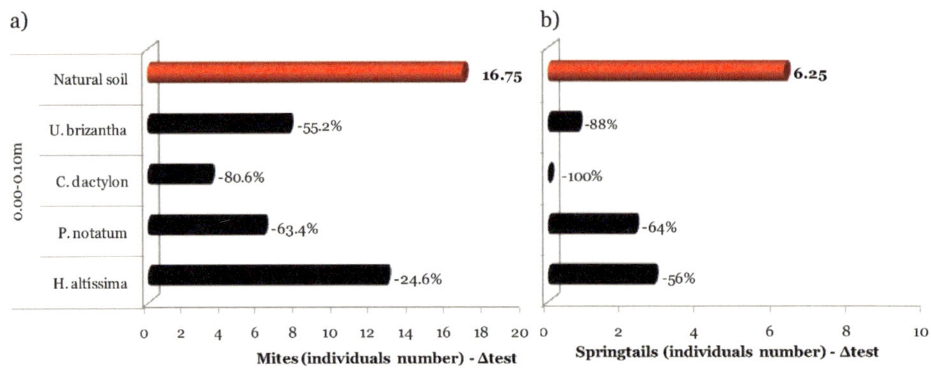

Figure 10.
Differences (Δtest) between the mites (a) and springtails population (b) of minesoil after 10.9 years of revegetation (under perennial grasses) relative to natural soil (under native vegetation).

high energy expenditure in maintaining it with greater respiration and CO_2 release rather than less carbon uptake into microbial cells.

Regarding the edaphic mesofauna, after 10.9 years of revegetation, the minesoil had a smaller mite population (between -24.6 and -80.6%) and a smaller springtail population (between -56 and -100%) compared to the reference soil. (**Figure 10a,b**), which was the consequence of the degraded state of the minesoil. On the other hand, it was observed that mites were larger than springtails population in both constructed minesoil and natural soil. This result is coherent because mites occur more in the interior of the soil, while the springtails occur on the surface [26].

According to the research results, it can be seen that the recovery of minesoils was more effective after 8.6 years of revegetation only in the physical condition up to 0.10 m depth. However, all the soil physical attributes and organic matter content are still far from the levels observed in the natural soil. The use of species with a more aggressive root system, such as the species selected in the present study (perennial grasses), possibly contributed to the positive results obtained in the short term, while it is expected that a following similar period (i.e., mid-term) is necessary for improvements in physical attributes below the 0.10 m layer.

About biological attributes, the 10.9 years of revegetation have not been sufficient yet to restore a mites and springtails population close to the natural soil.

4. Final considerations

Research results show that the reclaimed soils properties in coal mining areas, even after several years of reclamation, are still evolving and behind the quality of natural soils, especially the physical properties. This means that the reclaimed soil after mine decommissioning will be probably more fragile under cultivation than the natural soils, implying farmers to increase soils conservationist care in the first years. It is advisable that mining companies be aware of this and recommend farmers to cultivate the reclaimed soils using conservation systems, like no tillage systems, always maintaining straw covering on the soil's surface.

Acknowledgements

The authors would like to acknowledge Companhia Riograndense de Mineração (CRM), Brazilian Coal Network, CAPES e CNPq for logistical and financial support.

Conflict of interest

The manuscript is original, has not been published before, and is not being considered for publication elsewhere in its final form neither in printed nor in electronic format and does not present any kind of conflict of interests. The publication has been approved by all coauthors as well as by the responsible authorities at the institute where the work has been carried out.

Author details

Luiz Fernando Spinelli Pinto[1], Lizete Stumpf[1*], Pablo Miguel[1], Leonir Aldrighi Dutra Junior[2], Jeferson Diego Leidemer[2], Lucas da Silva Barbosa[2] and Mauricio Silva e Oliveira[1]

1 Agronomy College, Federal University of Pelotas, Brazil

2 Soil and Water Management and Conservation Program, Federal University of Pelotas, Brazil

*Address all correspondence to: zete.stumpf@gmail.com

References

[1] IEA—International Energy Agency. Coal 2019, Analysis and Forecasts to 2024. Fuel Report. 2019. Available from: https://www.iea.org/reports/coal-2019

[2] CGEE—Centro de Gestão e Estudos Estratégicos. Roadmap Tecnológico para Produção, uso Limpo e Eficiente do Carvão Mineral Nacional: 2012 a 2035. 2013. Available from: www.cgee.org.br/atividades/redirect/7877

[3] CRM—Companhia Riograndense de mineração. Relatório de Gestão 2015-2018. 2019. Available from: http://www.crm.rs.gov.br/conteudo/864/?Carvao-Mineral#.XuUbGm5FzIV

[4] Stumpf L. Desenvolvimento radicular de gramíneas perenes e sua eficiência na recuperação de atributos físicos de um solo construído compactado em área de mineração de carvão [thesis]. Pelotas: Universidade Federal de Pelotas; 2015

[5] Howard JL, Daniels WL. Soils of urban and human-impacted landscapes. In: Schaetzl RL, editor. International Encyclopedia of Geography: People, the Earth, Environment and Technology. New York: Wiley–AAG International Encyclopedia of Geography; 2018. pp. 1-11. DOI: 10.1002/9781118786352.wbieg0147.pub2

[6] WRB. World Reference Base for Soil Resources 2014: International Soil Classification System for Naming Soils and Creating Legends for Soil Maps. Rome: FAO; 2014. p. 192. (World Soil Resources Reports 106)

[7] Sencindiver JC, Ammons JT. Minesoil genesis and classification. In: Barnhisel RI, Darmody RG, Daniels WL, editors. Reclamation of Drastically Disturbed Lands. Madison: ASA, CSSA, SSSA; 2000. pp. 595-613. (Agronomy, 41)

[8] Daniels WL, Haering KC, Galbraith JM. Mine soil morphology and properties in pre- and post-SMCRA coal mined landscapes in southwest Virginia. In: Proceedings of the America Society of Mining and Reclamation. Lexington: ASMR; 2004. pp. 421-449. DOI: 10.21000/JASMR04010421

[9] Sobek A. Field and laboratory methods applicable to overburdens and minesoils. Virginia: Industrial Environmental Research Laboratory, Office of Research and Development, US Environmental Protection Agency. 1978

[10] Sobek A, Skousen JG, Fisher SE. Chemical and physical properties of overburden and minesoils. In: Barnhisel RI, Darmody RG, Daniels WL, editors. Reclamation of Drastically Disturbed Lands. Madison: ASA, CSSA, SSSA; 2000. pp. 77-104. (Agronomy, 41)

[11] Zhao Z, Shahrour I, Bai Z, Fan W, Feng L, Li H. Soils development in opencast coal mine spoils reclaimed for 1-13 years in the west-northern loess plateau of China. European Journal of Soil Biology. 2013;**55**:40-46

[12] Ahirwal J, Maiti SK. Assessment of soil properties of different land uses generated due to surface coal mining activities in tropical Sal (*Shorea robusta*) forest, India. Catena. 2016;**140**:155-163. DOI: 10.1016/j.catena.2016.01.028

[13] Feng Y, Wang J, Bai Z, Reading L. Effects of surface coal mining and land reclamation on soil properties: A review. Earth Science. 2019;**191**:12-25. DOI: 10.1016/j.earscirev.2019.02.015

[14] Haigh MJ, Sansom B. Soil compaction, runoff and erosion on reclaimed coal-lands (UK). International Journal of Mining, Reclamation and Environment. 1999;**13**:135-146. DOI: 10.1080/09208119908944239

[15] Mosebi PE, Truter WF, Madakadze IC. Smuts finger grass (*Digitaria Eriantha* cv Irene) root growth assessment and some physicochemical characteristics on coal mined land compacted soil. Journal of Applied Sciences and Environmental Management. 2018;**22**:1293-1296. DOI: 10.4314/jasem.v22i8.24

[16] Krümmelbein J, Raab T. Development of soil physical parameters in agricultural reclamation after brown coal mining within the first four years. Soil and Tillage Research. 2012;**125**:109-115. DOI: 10.1016/j.still.2012.06.013

[17] Angel HZ, Stovall JP, Williams HM, Farrish KW, Oswald BP, Young JL. Surface and subsurface tillage effects on mine soil properties and vegetative response. Soil Science Society of America Journal. 2018;**82**:475-482. DOI: 10.2136/sssaj2017.09.0329

[18] Stumpf L, dos Anjos LO, Pauletto EA, Pinto LFS. Roots of perennial grasses in the recovery of soils degraded by coal mining in southern Brazil. In: Tadele Z, editor. Grasses as Food and Feed. London: IntechOpen; 2018

[19] Wick AF, Daniels WL, Nash WL, Burger JA. Soil aggregation, organic matter and microbial dynamics under different amendments after 27 years of mine soil development. In: Proceeding of National Meeting of the American Society of Mining and Reclamation. Lexington: ASMR; 2010

[20] Mukhopadhyay S, Maiti SK, Masto RE. Use of reclaimed mine soil index (RMSI) for screening of tree species for reclamation of coal mine degraded land. Ecological Engineering. 2013;**57**:133-142. DOI: 10.1016/j.ecoleng.2013.04.017

[21] Bohrer SL, Limb RF, Daigh AL, Volk JM. Belowground attributes on reclaimed surface mine lands over a 40-year chronosequence. Land Degradation and Development. 2017;**28**:2290-2297. DOI: 10.1002/ldr.2758

[22] Akala VA, Lal R. Soil organic carbon pools and sequestration rates in reclaimed minesoils in Ohio. Journal of Environmental Quality. 2001;**30**:2098-2104. DOI: 10.2134/jeq2001.2098

[23] Clayton HG, Wick AF, Daniels WL. Microbial biomass in reclaimed soils following coal mining in Virginia. In: National Meeting of the American Society of Mining and Reclamation. 2009

[24] Lei H, Peng Z, Yigang H, Yang Z. Vegetation succession and soil infiltration characteristics under different aged refuse dumps at the Heidaigou opencast coal mine. Global Ecology and Conservation. 2015;**4**:255-263. DOI: 10.1016/j.gecco.2015.07.006

[25] Zhang L, Jinmanwang W, Bai Z, Chunjuan LV. Effects of vegetation on runoff and soil erosion on reclaimed land in an opencast coal-mine dump in a loess area. Catena. 2015;**128**:44-53. DOI: 10.1016/j.catena.2015.01.016

[26] Li J, Zhou X, Yan J, Li H. Effects of regenerating vegetation on soil enzyme activity and microbial structure in reclaimed soils on a surface coal mine site. Applied Soil Ecology. 2015;**87**:56-62. DOI: 10.1016/j.apsoil.2014.11.010

[27] Ahirwal J, Kumar A, Pietrzykowski M, Maiti SK. Reclamation of coal mine spoil and its effect on Technosol quality and carbon sequestration: A case study from India. Environmental Science and Pollution Research. 2018;**25**:27992-28003. DOI: 10.1007/s11356-018-2789-1

[28] Torroba-Balmori P, Zaldívar P, Alday JG, Fernández-Santos B, Martínez-Ruiz C. Recovering Quercus species on reclaimed coal wastes

using native shrubs as restoration nurse plants. Ecology Engineering. 2015;**77**:146-1533. DOI: 10.1016/j.ecoleng.2015.01.024

[29] Onweremadu EU. Chronosequential Pedon development on a mined landscape. Journal of American Science. 2007;**3**:16-22

[30] Nunes MCD. Condições físicas de solos construídos na área de mineração de carvão de Candiota-RS [dissertation]. Universidade. Pelotas: Federal de Pelotas; 2002

[31] Franco AMP. Caracterização física de um solo construído na área de mineração de carvão de Candiota-RS [dissertation]. Universidade Federal de Pelotas; 2006

[32] Gonçalves FC. Efeito de plantas de cobertura sobre os atributos físicos de um solo construído na área de mineração de carvão de Candiota-RS após três anos [dissertation]. Pelotas: Universidade Federal de Pelotas; 2008

[33] Miola ECC. Qualidade física de um solo construído e cultivado com diferentes plantas de cobertura na área de mineração de Candiota—RS [dissertation]. Pelotas: Universidade Federal de Pelotas; 2010

[34] Leal OA. Frações e qualidade da matéria orgânica de um solo construído vegetado com gramíneas após a mineração de carvão [dissertation]. Pelotas: Universidade Federal de Pelotas; 2011

[35] Catro RC. Avaliação temporal de atributos físicos de um solo construído em área de mineração de carvão recuperado com gramíneas perenes [dissertation]. Pelotas: Universidade Federal de Pelotas; 2012

[36] Stumpf L, Pauletto EA, Pinto LFS. Soil aggregation and root growth of perennial grasses in a constructed clay minesoil. Soil and Tillage Research. 2016;**161**:71-78. DOI: 10.1016/j.still.2016.03.005

[37] Gee GW, Bauder JW. Particle-size analysis. In: Klute A, editor. Methods of Soil Analysis. Madison: American Society of Agronomy; 1986. pp. 383-411

[38] Imhoff S, Silva AP, Dexter AR. Factors contributing to the tensile strength and friability of Oxisols. Soil Science Society of America Journal. 2002;**66**:1656-1661. DOI: 10.2136/sssaj2002.1656

[39] Dexter AR, Kroesbergen B. Methodology for determination of tensile strength of soil aggregates. Journal of Agricultural Engineering Research. 1985;**31**:139-147. DOI: 10.1016/0021-8634(85)90066-6

[40] Kemper WD, Rosenau RC. Aggregate stability and size distribution. In: Klute A, editor. Methods of Soil Analysis, Agronomy. Madison: American Society of Agronomy and Soil Science Society of America; 1986. pp. 425-442

[41] Palmeira PRT, Pauletto EA, Teixeira CFA, Gomes AS, Silva JB. Soil aggregation of an Albaqualf submitted to different soil tillage systems. Revista Brasileira de Ciência do Solo. 1999;**23**:189-195

[42] Embrapa (The Brazilian Agricultural Research Corporation) – Centro Nacional de Pesquisa de Solos (Soils Research Center). Manual de Métodos de Análise de Solo. Rio de Janeiro: EMBRAPA-CNPS; 2011

[43] Tedesco MJ, Bissani CA, Bohnen H, Volkweiss SJ. Análises de solo, plantas e outros materiais. Porto Alegre: Faculdade de Agronomia; 1995. p. 174

[44] Anderson JPE, Domsch KHA. Physiological method for the quantitative measurement of microbial

biomass in soil. Soil Biology and Biochemistry. 1978;**10**:215-221. DOI: 10.1016/0038-0717(78)90099-8

[45] Bachelier G. La faune des sols, son écologieet son action. Orston, Paris: Initiations at Documents Techniques, 38; 1978

[46] Reis DA, Lima CLR, Pauletto EA. Tensile strength of aggregates and compressibility of a soil built up with cover crops in a coal mining area in Candiota, RS, Brazil. Revista Brasileira de Ciência do Solo. 2014;**38**:669-678. DOI: 10.1590/S0100-06832014000200031

[47] Daniels WL, Zipper CE. Creation and Management of Productive Mine Soils. Powell River Project Reclamation Guidelines for Surface-Mined Land in Southwest Virginia. Virginia: Virginia Polytechnic Institute and State University; 2010

[48] Wick AF, Daniels WL, Nash WL, Burger JA. Soil aggregation, organic matter and microbial dynamics under different amendments after 27 years of mine soil development. In: National Meeting of the American Society of Mining and Reclamation. Pittsburgh/Lexington: ASMR; 2010. pp. 364-1386

[49] Anderson JD, Ingram LJ, Stahl PD. Influence of reclamation management practices on microbial biomass carbon and soil organic carbon accumulation in semiarid mined lands of Wyoming. Applied Soil Ecology. 2008;**40**:387-397

Section 2

Mining Techniques - Future

Chapter 5

Polish Experience in Shaft Deepening and Mining Shaft Hoist Elongation

Paweł Kamiński

Abstract

Deepening of active mine shaft comprises number of specific an very difficult operations, because it calls for use of untypical devices securing hoist operation in the shaft, as well as special technology tailored to actual technology of the deepened shaft face. The Leon IV shaft at the Rydułtowy mine is one of the last mining shafts deepened in Polish coal mining from the surface, and then deepened and finally equipped with mine shaft hoist installation. This investment will allow for the construction of the exploitation level of 1150 m and the availability of further coal extraction up to a depth of 1200 m. It will guarantee the possibility of exploitation of over 65 million tons of coal and continuous operation of the mine until 2040. At the same time, for the first time in the Polish hard coal mining industry, a flexible guiding of the mining cage and skips was used, which in comparison with rigid guiding is a much cheaper solution and has many other advantages. The chapter presents most important problems and technical solution implemented during construction and deepening of the Leon IV Shaft at Rydułtowy Coal Mine in Poland.

Keywords: shaft hoist elongation, deepening mining shafts, shaft construction, adjustable guiding

1. Introduction

Hard coal mine Rydułtowy is one of the oldest Polish mines in Rybnik Coal District. Its predecessor named Charlotte started production in the year 1806 and it was one of the greatest mines at that time. The mine as the first was equipped with steam engine already in the year 1855 and it was connected with the rest of the country via railway line what facilitated coal sales. At the beginning of twentieth century, the mine in question passed through numerous crisis phases that resulted in the employment reductions, and in the year 1932 the mine was even closed for a period of 4 years. However, the mine developed during the Second World War—because Germans needed big amounts of good-quality coal. In the period 1940–1944, the employment was increased threefold up to 3582 workers. After the Second World War, the Charlotta mine was renamed as Rydułtowy mine, which belonged to various structures of Polish mining industry. In the year, the KWK Rydułtowy was joined with Anna mine forming mining plant named as KWK Rydułtowy-Anna.

In the period 1990–1998 a new shaft Leon IV was sunked with diameter 8.5 m, which rarely occured in Polish mining industry. In this period, it was decided that the shaft depth of 1076.2 m would allow development of the new level 1050 m, which would fully satisfy the mine operational needs. The large resource base of the mine at a depth of less than 1000 m and the need to avoid sub-level mining has become the basis for undertaking another investment consisting in deepening the Leon IV shaft to a depth of 1210.7 allowing development of next exploitation level 1150 m. In the year 2013, design works were started and the process of shaft pipe deepening and extension of two shaft hoists (main and auxiliary) to the depth of 1000 m have been started (https://vimeo.com/321070029).

Flexible-ropes guiding of the shaft hoist cages in the Leon IV were implemented for the first time in Polish hard coal mining industry. Shaft deepening and necessity of extension of shaft hoists to the depth of 1150 m constituted great challenge both for designers and the unit realizing building works. It should be noted that like in each active coal mine, deepening of the active shaft is related with necessity of its continuous and undisturbed operation. In case of such technological restriction, works related with shaft deepening call for special securities, among others leaving of the rock shelf called as natural bottom, or building in the shaft the so-called artificial bottom.

In such cases, transport works in deepened shaft section call for building of auxiliary hoist device with underground hoist machine of special turnstile adapter for personnel transport. Big-diameter hole used for transport, water drainage, and fresh air can greatly facilitate the works related with shaft deepening. However, in such cases, excavation on the level to which the shaft is deepened is needed. The shaft Leon IV can be a good example of application of new technical and technological solutions. Three of such solutions will be discussed in the present study:

- single-layer waterproof lining within the Section 782.0–932.0 m,
- shaft deepening technology within the Section 1076.2–1210.7 m, and
- extending of shaft hoists from the level 1000.0 (960 m) to the mining level 1050 m and auxiliary level 1200 m used for the mine water drainage.

2. Single-layer sulfate-proof lining

In original project of the shaft IV, two-layered lining with hydroinsulating shield made of PE foil was foreseen for the shaft Section 782,0–932,0 characterizing with occurrence of sulfate and magnesium waters. Such linings were commonly used by KOPEX—Shaft Building Company S.A. Sinking technology within the section in question foresees the following works [1]:

- between ordinates 784.5 and 786.0 m: building of the B15 class concrete lining (currently C12/15 concrete class),
- between ordinates 786.0 and 932.0 m: building of preliminary shaft lining (in direction from top to the bottom) in form of 0.56 m thick shaft concrete block wall,
- between ordinates 930.5 and 932.0 m: building of shaft brick made of B30 class concrete B30 (at present C25/30), and
- making of the inner layer of the final lining from concrete class B25 and B30 (at present C20/25 and C25/30) wet laid from the bottom up after laying on the walls and tight sealing of the 2 mm thick PE foil jacket.

Because at this time one of the Polish cement factories produced special Portland cement called as bridge portland cement CP 45(M) marked with symbol CP 45(M) resistant to strong sulfate and magnesium aggression, after research works executed in the AGH University of Science and Technology, modification of the construction and technology of completion of shaft lining section, has been proposed. New project of single-layered lining lied from the top to the bottom following the shaft face advance comprised making of 0.65-m-thick single-layered concrete lining lied in wet system. Due to the dependence of calculated pressure on the shaft lining, its bearing capacity was controlled due to concrete strength with use of two receipts [1] for concrete of class B25 marked as R25/1/2 and for concrete B30 marked as R30/1/4. All concretes were prepared on the basis of Bridge Portland cement CP 45(M).

Receipts developed in the AGH University of Science and Technology and verified by laboratory tests guaranteed suitable strength of the concrete and suitable bearing capacity of the lining of targeted thickness, as well as suitable water tightness of the level W8. Concrete lining was made in 4-m-long sections in direction from the top to the bottom. In such technology, waterproofness depends mainly on technological joints between upper (old) section and bottom (new) section. In the project in question, re-sealing of these joints was made first time in Polish shaft building with the use of injection hoses of the type FUKO 2 (**Figure 1**), with former opinion from Higher Mining Office concerning their security due to the presence of methane, after special tests were conducted in Experimental Mine Barbara in Poland. The injection hoses FUKO 2 were mounted to upper section of the shaft with use of metal connectors fitted with screwed joints. After the next lining, the section was concreted, the fissure was filled with a binding mixture on its whole length (see **Figures 1** and **2**) obtaining satisfactory sealing of the neuralgic element of the shaft lining.

Another issue that was solved during the construction of this shaft was the rock drainage system behind the lining. It is well known that water accumulation behind a waterproof lining is dangerous due to the possibility of high hydrostatic pressures appearing on the casing after joining various aquifers with a shaft.

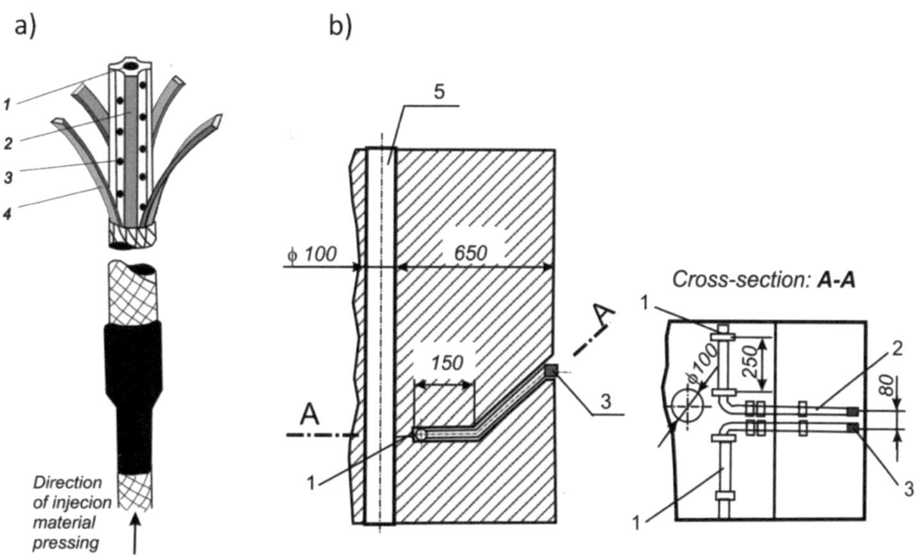

Figure 1.
Sealing system of the concrete shaft lining with use of hoses FUKO2 [1]. (a) Injection hose FUKO 2. Markings: 1—injection channel Φ = 10 mm; 2—hose core; 3—injection holes; and 4—neoprene ribbons playing role of non-return valves. (b) Housing of hoses in technological joint of sequent sections of the concrete shaft lining. Markings: 1—injection hose FUKO 2; 2—steel pipes; 3—threaded ending for pressure hoses; 4—technological joint between two concrete lining section; and 5—drainage pipeline for the rock body dewatering.

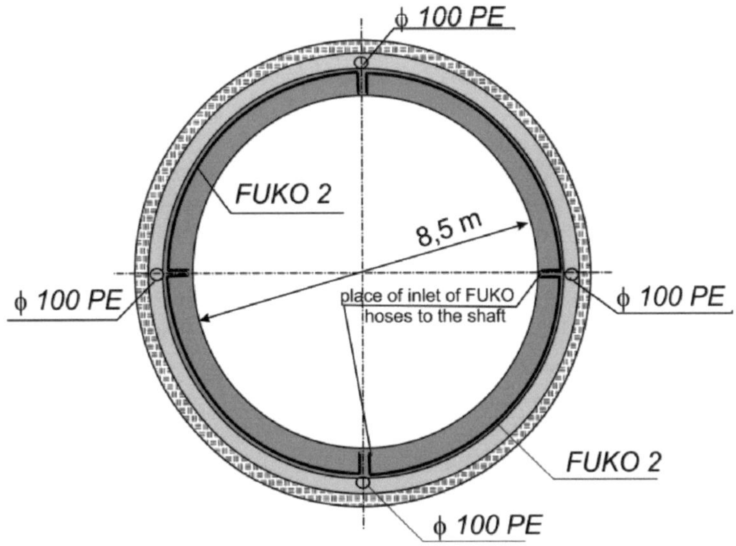

Figure 2.
Injection system via hoses FUKO 2 and rock body drainage system behind the shaft lining [1].

This problem was solved by laying vertical drainage 100 mm diameter pipelines on four azimuths along the lining arranged (**Figure 2**).

3. Chosen elements of the shaft Leon IV deepening

This, about 80-million PLN, investment in the Leon IV shaft deepening by next 140 m resulted from the necessity of the production processes' modification in the Rydułtowy mine [2]. This modification comprised first of all of shortening of the time of personnel transport to the shift face, as well as facilitation of the needed materials' delivery and considerable improvement of the ventilation of this part of the mine.

The investment task related with development of mine infrastructure in Leon IV region comprised the following activities:

- technical designs,
- physical shaft deepening and reinforcing,
- making the two-way inlet to pit bottom of exploitation level at the depth of 150 m,
- making the inlet to single-way pit bottom at the depth of 1200 m destined for needs of the mine main drainage system.
- elongation of mining hoists: main to the level of 1150 m and auxiliary to the depth of 1200 m, and
- installation of needed elements of mechanical equipment of the inlets to pit bottoms of both built levels.

Targeted depth of 1210.7 m was reached in August 2016. After completion of the reinforcing of the deepened shaft part in the year 2017, pioneering works in Polish mining industry works related with elongation of shaft hoists, including untypical and difficult elongation of guiding hoists accompanied by exchange of all leading ropes, have been executed (**Figure 3**).

3.1 Technology of the Leon IV shaft deepening

The Leon IV shaft was deepened to level 960 m, keeping full exploitation ability. Works related with shaft deepening were conducted with artificial bottom of

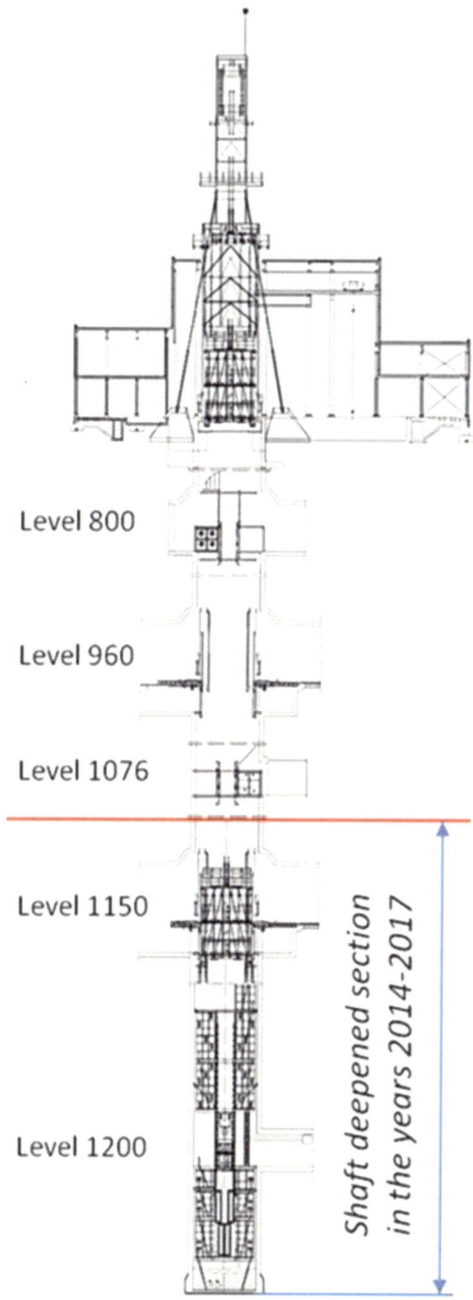

Figure 3.
Leon IV shaft profile [2].

special construction of two platforms joined with vertical partition. Mining works were conducted by standard method with use of explosives. Because haulage is the greatest problem in deepened shafts, in the case in question, at first a dike was made on level 1200 m, in order to make great diameter hole of the length of 115 m reaching the shaft bottom before deepening. The hole was located in such manner that its axis was located at a distance of 2.2 m toward east from the shaft axis, which allowed collision-free operation of the basket covering the hole inlet in the shaft face. The 1200 mm diameter hole was nevertheless endangered by the possibility of rock blockage. In order to remove the jam in the hole, a rope of 25 mm diameter with conveyor scrappers was installed in the hole. The rope vertical movements stimulated by low-speed winches KUBA-5 installed in ditches on the level 1076 and 1200 m, (see **Figure 4**) provoke fall down and the hole clearance.

Transport in the whole deepening process was handled by special devices located in ditch on level 1076 m (**Figure 4**):

- hoist machine B-1500 for bucket handling,
- two low-speed winches KUBA-10 for adjustable formwork handling,
- windlass KUBA-5 for rope handling,
- windlass KCH-9 for basket hanging protecting hole in shaft face, and
- supporting construction for assemblage of wheels during shaft deepening.

In the ditch, at the level 1200 m, low-speed windlass KUBA-5 with track wheel for the hole clearing rope (see **Figure 4**) was installed. Single–layered lining of C30/37 concrete lied in wet system with use of steel moveable formwork of the height 2.15 m, has been applied. Calculated and consulted with the Investor lining has thickness from 0.5 to 0.6 m. With respect to expected small and ephemeral water inflows into the shaft, no special waterproof precautions were designed [4].

3.2 Start-up of levels 1150 and 1200 m

Thanks to the Ruch Rydułtowy investment, it will allow the exploitation from coal seams No 713/1–2 and 712/1–2, which belong to the most promising mining assets within mining areas belonging to this part of Rybnik Mining District. Development of this part of the deposit will allow building the new level at the depth of 1150 m (https://vimeo.com/320940852). Two-way inlet is equipped with full set of the wheel transport handling, with special platform for material reloading from wheel into lifted gondola transport. The inlet performances are as follows:

- excavation founding depth: 1143.7 m,
- height: 7.3 m,
- width: W side—8.11 m and E side—7.0 m, and
- basement depth: 2.30 m.

The pit-bottom geometry with use of 3D visualization is shown in **Figure 5**. Universality is characteristic feature of the shaft pit–bottom 1150 m—main transport level (https://vimeo.com/333420960). Within this level there is a possibility of using three shaft hoists, what will considerably accelerate process of material lifting, as well as it will allow fast and fluent personnel transport. Using transport platforms forced equipping shaft pit-bottom basement with devices and machines needed for pushing mine trucks into large-size mining cages, as well as into standard cages. The deepening of the Leon IV shaft was also used to reorganize water management in this area. For this purpose, excavations needed for the main water drainage handling were localized in one-way pit-bottom at the level 1200 m, and the elongation of this level to auxiliary hoist was necessary, and it was requalified from auxiliary hoist into "small" hoist.

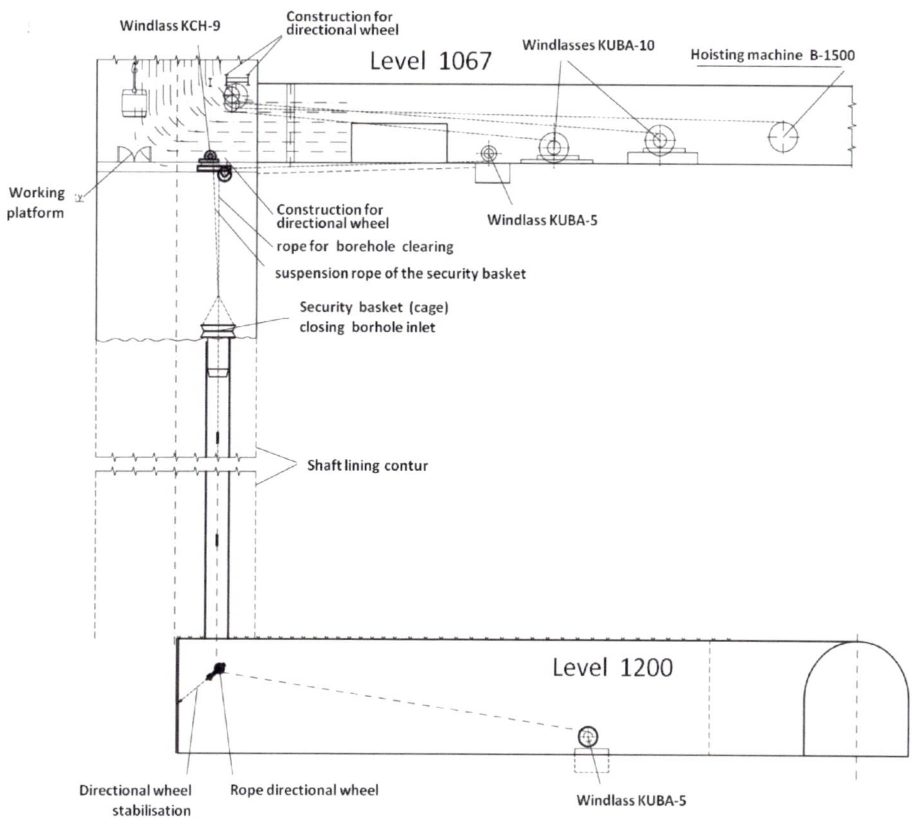

Figure 4.
Distribution of devices during Leon IV shaft deepening at levels 1076 and 1200 m [2].

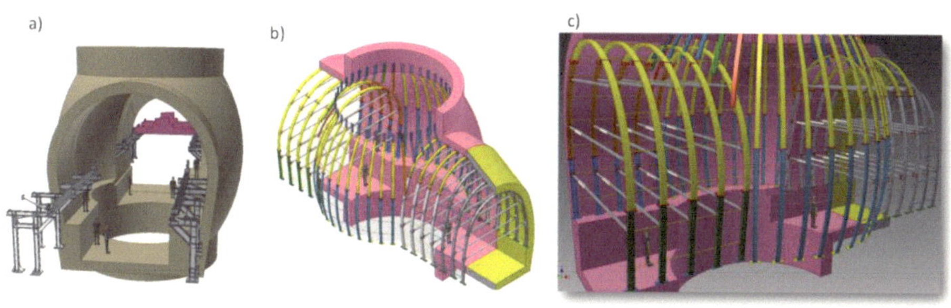

Figure 5.
3D model of the pit-bottom—level 1150 m [3]. (a) General view; and (b) and (c) 3D model of the steel-concrete pit-bottom lining.

Geometry of the shaft inlet at level 1200 m:

- depth of the excavation founding: 1195.7 m,
- height: 4.2 m, and
- width: 6.1 m.

The shaft inlet has anchor-concrete-steel lining and is equipped with level guidance construction with oscillatory platform in the inlet basement.

3.3 Furnishing of the shaft Leon IV

As the main transport shaft, the shaft Leon IV is equipped with three compartments: one for main hoist with large-size cage, second for ordinary three-deck cage, and auxiliary hoist. The shaft cages are suspended on two 48-mm-diameter rope carriers driven by drive wheel Koeppe. In order to balance masses of rope carriers, two equalizing ropes of diameter Φ = 53 mm are installed.

Shaft Leon IV is the first shaft in Polish hard coal mining industry, in which flexible guiding of shaft cages or skips has been extended. Guiding and defender ropes are suspended on wedge-shaped spreader beams located over beams of the shaft tower. The guiding and defender ropes hang down freely and are tensed by the attached weights of such mass that each 100 m of the shaft depth corresponds to tensing power of the value at least 8 kN. The guiding and defender ropes are mounted in special baskets located below lower guiding rope

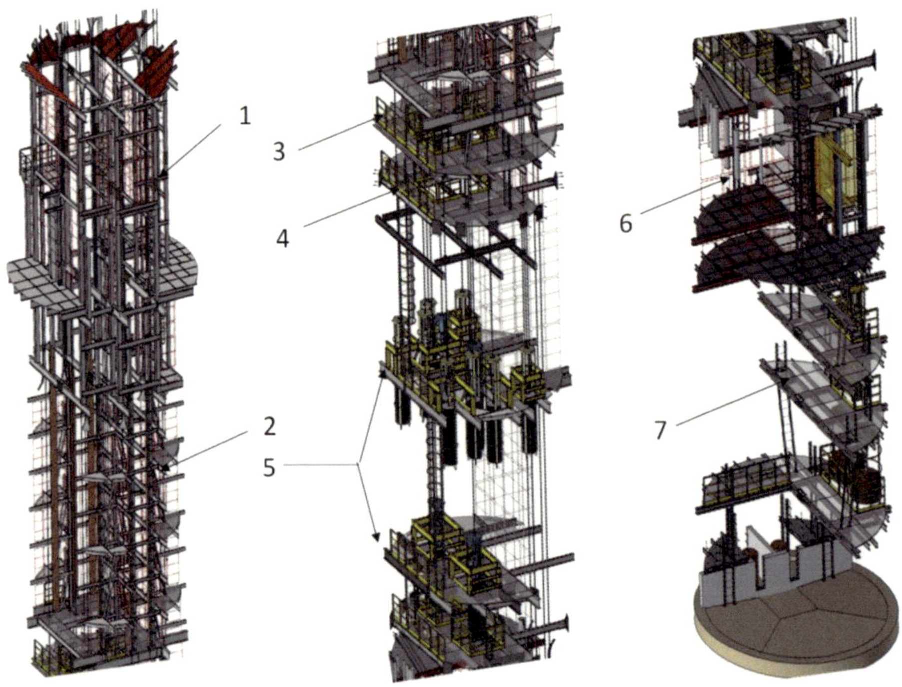

Figure 6.
Model of the Leon IV shaft furniture after deepening to the depth of 1210 m [2]. Markings: 1—level guidance at level 1150 m; 2—breaking system of the main shaft hoist; 3—return station of the equalizing ropes; 4—control platform of the equalizing ropes; 5—control platforms of guiding rope weights; 6—level guidance at level 1200 m; and 7—furniture of the Leon IV shaft sump.

frame. In case of the Leon IV shaft, the shaft furniture consists of the following elements: (**Figures 6** and 7).

Elastic guiding of the shaft cages comprises 12 guiding ropes, 4 defender ropes, and 3 rope carriers and equalizing ropes between large-size cage and three-deck cage (**Figure** 7).

Profits resulting from application of rope guiding of hoist cages or skips are great. The profits are as follows:

- low cost of used materials,
- easy handling,
- long durability,
- short assembling time in new shaft,
- guiding of the cages ore skips in the shaft is soft, without of tremors and side hits,
- possibility of fast shaft hoist operation,
- quiet run cages ore skips results in elongation of the rope carrier, and
- ventilation resistances are almost 10 times lower than in case of shafts with rigid guides.

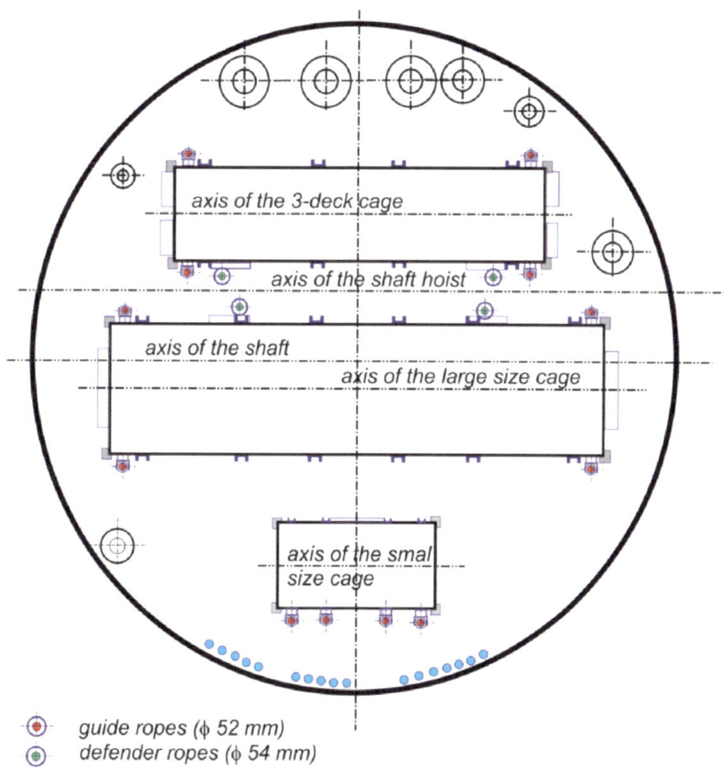

Figure 7.
Elements of the elastic guiding of shaft Leon IV.

Linear guiding of cages has also some disadvantages, like:

- possibility of transverse movements, which are vertical to the running direction, which calls for longer operation intervals than in case of rigid guides,
- serious difficulties in exploitation result—rock body movements,
- need of great size shaft diameters and difficulties related with relocation of shaft hoist devices in numerous transport sections,
- tensions of guiding ropes reaching values of 2000 kN causing additional loading of the shaft tower and application of suitable construction needed.
- rope tensioning devices require a properly managed sump,
- in case of application of two-cage hoists, the application of additional ropes called as "defender ropes" protecting against collisions of vessels moving in two different directions is needed.
- rope carrier must be made of non-rotating rope, and
- the main advantages of the application of such type of guiding result from the fact that the use of tens shaft fastening frames is not necessary.
- the main benefits of using this type of guide arise from the fact that the shaft does not require the installation of dozens of rigid guides frames.

The other element of the shaft furniture is related with main hoist cages' braking system localized under level 1150 m and in the shaft tower. This system is composed of thickened wood guides. The other elements of the shaft furniture comprise:

- platform of the return station of equalizing rope,
- platform of equalizing ropes control system,
- positioning frame of weighs of guiding ropes and bumper ropes,
- control platforms of guiding and bumper ropes, and
- protective platform.

Necessity of equipping the flexible guidance with special corner guiding for shaft vessels is essential element of guiding on individual levels. In case of the shaft Leon IV it refers to levels 800, 1067, 1150, and 1200 m.

3.4 Elongation of mine shaft hoists

Shaft deepening is strictly related with necessity of mine shaft hoists. In case of the Leon IV shaft, in construction of elastic guiding of hoist cages to level 1067 m, the ropes longer than the exploited shaft were applied. Excess of ropes was stored on special drums located in pit-bottom of the level 1076 m. Thus elongation of these ropes required only suitable control of the rope destruction degree and then lowering them to the level 1150 m. Works related with elongation of the mine shaft hoist were started from auxiliary hoist and then lowering two guiding ropes of diameter Ø = 32 mm to level 1200 m. Weight of the rope of diameter 32 mm amounts for around 5.3 ton, whereas for ropes of diameter Ø = 54 mm, this weight reaches value over 25 ton. Thus, every rope maneuvers, that is, their raising or requires execution of assembling works with use of special hoist machine having high lifting capacity, as well as special guiding wheels mounted on special platform. Low-speed lifting machine EWP-35 was used for rope lifting and lowering. After making welded clamps and taking rope weight by lifting machine, disassembling was made in order to check condition of ropes by suitable expert. After acceptance of the ropes for exploitation, they were lowered to level 1200 m, where weighting baskets were mounted. Next stage of elongation of the shaft hoist comprised elongation of guiding and bumper ropes of main hoist, that is, unwrapping rope reserve accumulated at the level 1067 m. After installation of additional constructions of rope wheels on the platform on lower rope wheel of the Leon IV shaft, that is, directing technological rope of Ø = 40 mm in place of wedge-shaped spreader beam, stage of basal works related with taking and lowering ropes in targeted place have been started. When the welded clamps were installed and the weight was taken by low-speed lifting machine (40 ton), the rope was lifted to level of the foundation in order to be checked by the expert, and then the rope was lowered again to the level of weights control platform. After completion of the operations of the first rope, the guiding wheel was relocated on the platform in such a manner that operations of the next ropes would be possible. This operation was repeated eight times for guiding lines of the main hoist cages, and four times for bumper ropes located between both vessels of the same hoist. Scheme of devices needed for hoist elongation and visualization of the guiding wheels on the tower are shown in **Figure 8**.

Technology of the rope carriers and equalizing ropes in the shaft Leon IV is similar to standard technologies of ropes operated in mine shafts. In this case, at first, some preparatory works related with construction of foundation of the lift EPR-1000 and installation of sheave wheels have been executed in the shaft foundation. Carrying ropes were lowered to the shaft after placing shaft cages on special platform and taking the weight of ropes by portable lift EPR-1000.

After relocation of the great-size cage to the level 1150 m and installation of spreader beams, the equalizing ropes were elongated. In this case, the rising

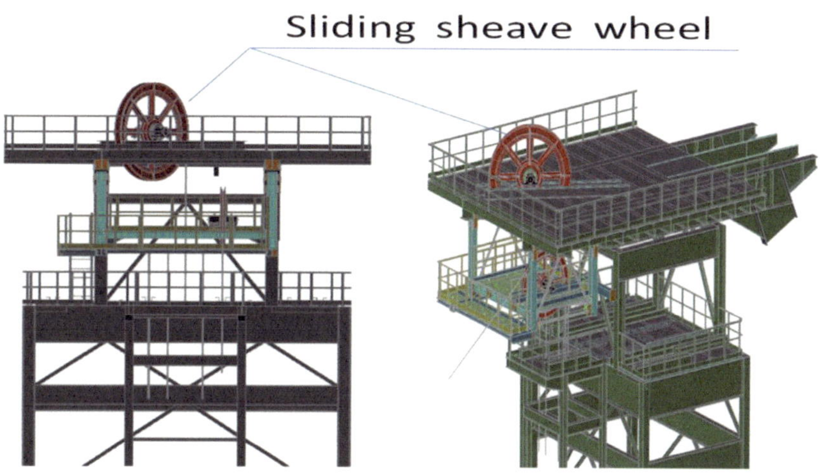

Figure 8.
Visualization of sheave wheel location on hoist tower [2].

large-size cage pulled new equalizing ropes. After relocation of three floor cages to the level 1150 m, construction of equalizing ropes under foot of three floor cage was completed.

3.5 Stabilization of the cage at the large-size Leon IV shaft hoisting system

Adjustable guiding system replaced the chairing mechanism of a rope guided cage at level 960 of "Leon IV" shaft. The mechanical part of the system is included in the project (**Figure 9**), developed to solve the issue of leading the cage through level 960. The most important factor taken into consideration was safety. With chairing, it could be ensured only at the expense of significant lengthening of the duration of a single cage ride from the ground level to level 1150 m and vice versa, since the velocity of the cage had to be reduced from 10 to 0.5 m/s, already 100 m before level 960 m. The changes between systems included replacing the angular guides with adjustable guides (including two pairs of upper and two pairs of lower guides) and the main support with four permanent frames, fixed to the shaft lining using additional girders, structurally independent from the construction of the chairing. In this arrangement, the adjustable guides can be switched between idle mode and working mode. The motion is restricted by:

- Part of a respective frame, known as roadway, serving as adjusting track for each pair of lower adjustable guides,
- Two articulated links (upper and lower), for each pair of upper adjustable guides,

Two pairs of lower adjustable guides are powered by a single hydraulic cylinder each, one end articulated to the guiding frame, the other to the guides itself. Each pair of upper guides is powered by two hydraulic cylinders, articulated with lower end to the frame, and upper end to the joint.

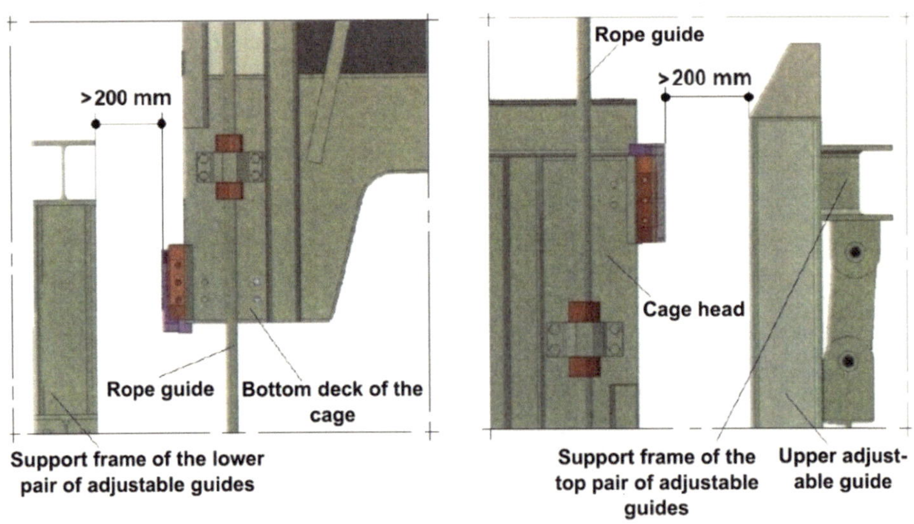

Figure 9.
System in idle mode.

During the hoisting operation (whether it is from the ground level to level 1150 m or in the opposite direction), each pair of the adjustable guides stays in idle mode (**Figure 9**).

This setup ensures safe movement of the cage through level 960 with velocity of 10 meters per second. Idle mode is also used when the cage moves between level 950 m and 1150 m (both directions). Setting to working mode, decided by the signalman at level 960, takes place once the cage stops at level 960. During the cycle each of the upper ends of the lower adjustable guides is inserted between pads of the emergency braking slide guide, attached to the face of the bottom deck. The upper ends of the upper adjustable guides are inserted between pads of the emergency breaking slide guide attached to the face of the cage's head. The work mode setup of the bottom and top adjustable guides is shown in **Figure 10**.

Layout of the entire system is shown in **Figure 11**. The adjustable guides consist of 180 x 260 mm box beams made of C260 C-profiles. Frames made of HEB 260 are attached to two technological beams with M24 Hex bolts, class 8.8.

Static analysis was performed for the constructions. Assumptions and results are shown in the schemes below:

Static analysis—lower frame

Known variables:

- Transportation unit load: 200 [kN]
- Continuous load:

$$q = \frac{200\ [kN]}{1.654\ [m]} = 120.92\ \left[\frac{kN}{m}\right] \tag{1}$$

The assumed load rounded to the value of 121 kN/m

- Factor of safety

$$\frac{480}{62.87} = 7.63 \tag{2}$$

- Force from the hydraulic cylinder: 139.2 [kN] (**Figure 12**)

Static analysis—upper frame

Known variables:

- Transportation unit load: 200 [kN]
- Continuous load:

$$q = \frac{200\ [kN]}{1.654\ [m]} = 120.92\ \left[\frac{kN}{m}\right] \tag{3}$$

The assumed load rounded to the value of 121 kN/m

- Factor of safety

$$\frac{480}{75.09} = 6.392 \tag{4}$$

- Force from the hydraulic cylinder: 139.2 [kN] (**Figure 13**)

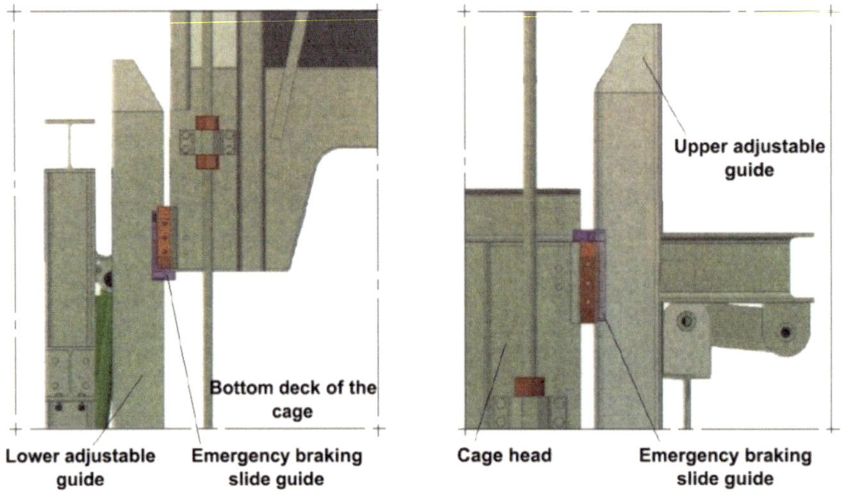

Figure 10.
System in working mode.

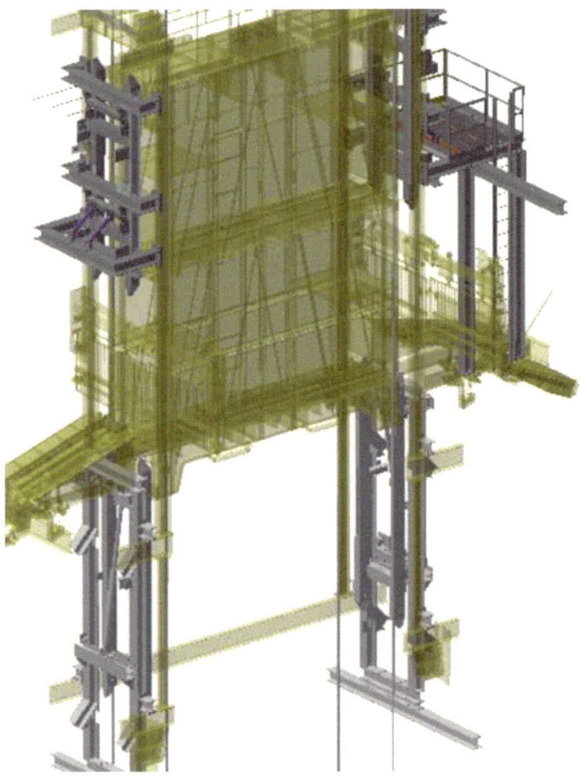

Figure 11.
Layout of the adjustable guiding (gray color).

Measurements

The main purpose of measurements was to determine:

- maximum forces applied to the head and the bottom deck of the cage during experimental cycles of loading and unloading the heaviest transportation unit approved for this type of transportation, performed at level 960,

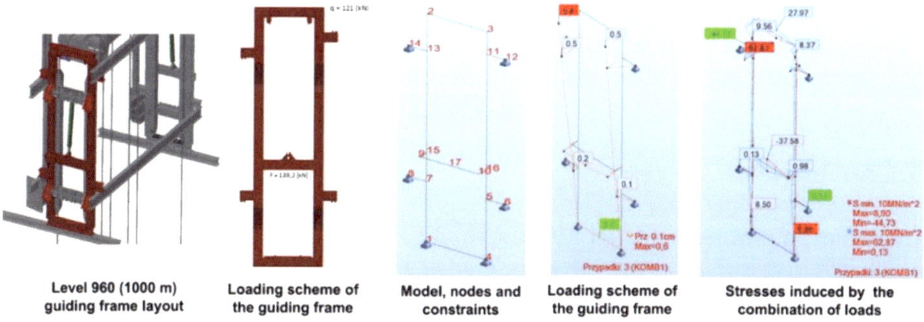

Figure 12.
Static analysis for the bottom frame.

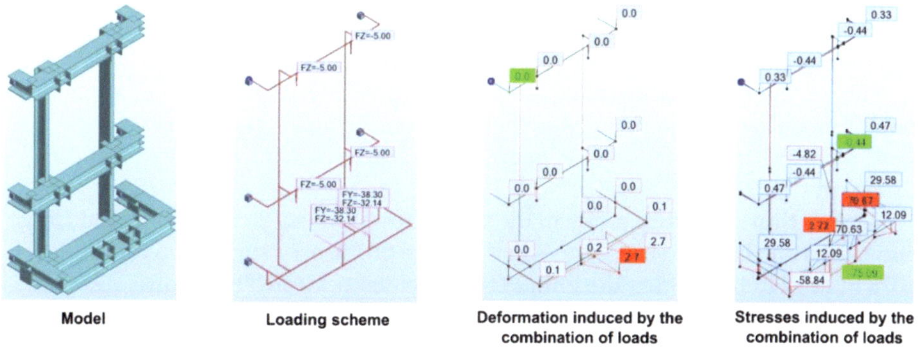

Figure 13.
Static analysis of the upper frame.

- maximum stresses induced in girders of the cage during experimental loading and unloading cycles, being the result of absorbing forces from the adjustable guides.

The force measurements were conducted accordingly to the previous ones, carried out on exactly the same cage in July 2018, also concerning the forces absorbed the cage during the experimental loading and unloading cycles, but forces from angular guides were replaced with forces from adjustable guides. Moreover, the scope did not include measuring the stresses induced in girders as a result of absorbing forces from angular guides, as these guides were also stabilizing the middle deck during the experiments.

The reason for adding stress measurement was the necessity to properly assess whether resigning from stabilization of the middle deck of the cage in this arrangement may be considered a viable assumption. Theoretical analysis based on calculation model from [5] demonstrated such possibility; however, empirical verification was deemed vital. It was implemented by equipping the main measurement unit with two external modules attached to the middle deck for the time of measurement operations. The cycles were divided in two stages:

- Stage one, covering the first six experimental loading and unloading cycles, concerned measuring the forces at the bottom deck of the cage as well as stresses induced in girders on the distance between the middle deck and the lower deck,

- Stage two, covering the following six experimental cycles, concerned measuring the forces at the top deck of the cage as well as stresses induced in girders on the distance between the middle deck and the head.

4. Conclusions

Technical problems related with the shaft Leon IV sinking and deepening presented in this chapter can be considered as an example of continuous innovation and development of the shaft building technology. Although nowadays shafts are rarely deepened, high level of modern technique and mechanization of both sinking and equipping the shafts indicate potential possibilities of further development of this building branch.

Application of new bridge cements M45 and modification of the philosophy of assuring the shaft lining tightness even during building the shaft Leon IV allowed implementation of very profitable replacement of multilayered lining with single-layer lining, which is less time-consuming and much cheaper.

Shaft deepening during its exploitation was possible only in result of application of modern construction of "artificial bottom," which tightly separated the shaft from the area of works conducted by the company named as Shaft Building Company S.A. (PBSz).

Application of elastic system in the shaft Leon IV in hard coal mining industry and use of much more longer ropes and storage of the rope surplus on the level 1078 m can be classified as the uniquely far-sighted project. This in turn allowed implementation of much more easy technologies of shaft hoists elongation.

Designed by Shaft Sinking Company, elongation of the shaft hoists, which was realized in possibly shortest stoppage of the shaft operation, was a pioneer and innovative venture. Total work comprised elongation of 20 ropes.

Modernization of the shaft Leon IV was a key element of the restructuring plan and development of the joint-venture mine ROW gathering mines: Jankowice, Chwałowice, Marcel, and Rydułtowy. Elongation of shaft hoists, development of main transport horizon at the level 1150 m, and development of the main drainage system at level 1200 m will allow considerable shortening of the time of personnel transport to exploitation excavations, which is related with considerable improvement and elongation of the personnel working time, that is, improvement of financial results of mine operation and whole mine ROW.

Conducted analysis and research confirm that the mechanical system, its attachment to the shaft lining and particular structural elements of the tested solution are correct and fulfill the conditions defined in the Decree of the Minister of Energy from November 23, 2016 concerning the detailed requirements of operating in underground mines (Journal of Laws of the Republic of Poland 2017, item 1118), as well as in the technical standard PN-G-46227: 2002—Mining shafts. Shaft equipment. Requirements.

Changes introduced at level 960 m (1000 m), made according to Annex No. 2, included removing the corner guides of the cage. The additional space obtained between the chairing elements ensures the safe passage of the cage through level 960 with the set velocity of 10 m/s, assuming the requirement of § 545 of the Decree of the Minister of Energy is met. After analyzing the results of measurements of the forces from bottom deck and the head of the cage absorbed by adjustable guides during loading and unloading cycles, it can be stated that replacing the corner guides with adjustable guides does not violate this approval.

Author details

Paweł Kamiński
AGH University of Science and Technology, Kraków, Poland

*Address all correspondence to: pkamin@agh.edu.pl

References

[1] Kostrz J, Olszewski J, Czaja P, Deja J, Witosiński J. Zastosowanie betonów odpornych na silną agresję siarczanową i magnezową w budownictwie podziemnym. Budownictwo Górnicze i Tunelowe. 2000;**3**:4-11. ISSN: 1234-5342

[2] Olszewski J, Czaja P, Bulenda P, Kamiński P. Pogłębianie oraz wydłużanie górniczych wyciągów szybowych—szybu Leon IV w kopalni KWK ROW—Ruch Rydułtowy. Przegląd Górniczy. 2018;**74**(8):7-16. ISSN: 0033-216X

[3] Czaja P, Kamiński P. Wybrane zagadnienia techniki i technologii głębienia szybów. Kraków: Szkoła Eksploatacji Podziemnej. 2016. p. 161. ISBN: 978-83-927920-9-3

[4] Kicki J, Sobczyk EJ, Kamiński P. Vertical and decline shaft sinking—Good practices in technique and technology. In: International Mining Forum 2015; 23-27 February 2015; Cracow, Poland. Leiden: CRC Press/Balkema; 2015. p. 197. ISBN: 978-1-138-02820-3

[5] Płachno M. Mathematical model of transverse vibration of a high-capacity mining skip due misalignment of the guiding tracks in the hoisting shaft. Archives of Mining Sciences (Archiwum Górnictwa). Wyd. Polska Akademia Nauk, Komitet Górnictwa, Kraków. 2018;**63**(1):5. DOI: 10.24425/118882

Chapter 6

Polish Experiences in Handling Water Hazards during Mine Shaft Sinking

Piotr Czaja, Paweł Kamiński and Artur Dyczko

Abstract

The geological structure of most Polish mining regions is rich in groundwater, making shaft sinking difficult. In recent years, more than a dozen shafts, some almost 700 m deep, have been sunk in Poland using various methods of water hazard elimination. The vast majority of shafts that pass through aquifer formations have been sunk using artificial rock freezing, waterproof tubing, and concrete lining. Generally, this system has proven to be very effective. However, there have been cases of complications during sinking, including occasional flooding. This paper presents two cases of highly problematic flooding in shaft sunk using the freezing method, both leading to considerable construction delays and a significant increase in shaft sinking costs. The first case involved water inflow into the bottom section of the R-XI shaft at KGHM with rocks near the melting point of ice. In the other case, problems occurred passing through an Albian layer in the S. 1.3 shaft sunk for the Lubelski Węgiel Bogdanka S.A. mining corporation, where the freezing process was carried out while it was necessary to heat the rocks in the upper part of the shaft to protect the final lining from damage.

Keywords: mining shaft, water hazard, grouting, dewatering

1. Introduction

Exposing deep-seated mineral deposits requires the construction of new shafts. In Poland, where usable minerals are usually covered by thick layers of heavily waterlogged overburden, the construction of new shafts poses extraordinary difficulties. New shafts continue to be designed and constructed in quite challenging hydrogeological conditions in Poland, as well as in other countries worldwide. Hence, it would be fruitful to look at some Polish experiences in coping with this extremely difficult hydrogeology while mining deposits of both hard-coal and nonferrous metal ores. A range of detailed examples of how to eliminate such water hazards has been provided elsewhere [1–3]. Over the last three decades, Poland has seen at least several cases involving shaft flooding. These occurred mainly during the sinking phase. There are many methods for eliminating water hazards and dewatering flooded shafts to put them back into operation. This paper presents two cases of highly problematic flooding in shaft sunk through highly waterlogged layers using the freezing method, both leading to considerable construction delays. The first case involved the removal of increased water inflow into the R-XI shaft at KGHM. In the other case, problems occurred due to a shaft passing through an Albian layer in the S. 1.3 shaft sunk for the Lubelski Węgiel Bogdanka S.A. mining corporation. Although completely different from each other, these cases provide useful guidance and a serious warning against hasty shaft design or a careless approach to constructing shafts [4–6]. Considered completely safe for shaft construction, the technological solutions presented here should be of interest to experts in water-related mining issues.

2. Diversion of increased water inflow into the R-XI shaft during sinking

Waterlogged overburden formations as deep as 700 m below the ground have made it necessary for Polish mining corporations to use the freezing method to construct all copper mine shafts and most hard-coal mine shafts. Hundreds of shafts have been successfully sunk in Poland using this technology. However, when it seemed that the engineers had virtually eliminated freezing pipe leaks in the boreholes, a major problem that had caused brine leaks into frozen rock, water hazards emerged in completely unexpected and highly unlikely situations.

2.1 Project specification and the effects of the water hazard

The R-XI shaft was not the first structure of this type constructed by PeBeKa S.A. in the Polish Copper Basin area [7]. Hydrogeological surveys preceding the shaft work at depths of 431.0–630.0 m indicated no significant water hazards along this section. The projected water inflows into the shaft face below the 431.0 m level are shown in **Table 1**. The R-XI shaft was designed to serve as a ventilation shaft and has the following parameters [5]:

- Lining diameter—7.5 m
- Total depth—1250 m
- Aquifer thill depth—630 m
- Freezing depth—635 m

At the time, this shaft had the greatest rock freezing depth at 635 m. PeBeKa Lubin applied many innovative rock freezing solutions. One of them was selective freezing using two types of freezing holes: short holes with a depth of 395 m and

Depth interval [m]	Water inflow [m^3/min]	
	Minimum to maximum	Average
431.0–460.0	0.042–0.070	0.056
460.0–470.0	0.042–0.070	0.056
470.0–500.0	0.061–0.330	0.160
500.0–565.0	0.205–0.490	0.334
565.0–630.0	0.334–0.550	0.425

Table 1.
Predicted water inflows into the shaft [7].

long holes with a depth of 635 m. This method made it possible to achieve a frozen mantle that was thickest in its lower portion, where the water pressure was found to be the highest.

Because a gallery had already been excavated near the shaft at a depth of 1212.7 m, the design included a simplified drainage system for the shaft face below the freezing zone. This was achieved through a dewatering borehole drilled in the shaft axis vertically upwards from a level of 1212.7 m. This made it possible to dispense with the construction of an expensive cascade drainage system and significantly facilitated shaft sinking at depths of 635–1212.7 m. At the 503.6–632.4 m shaft section, the design included a combined panel and concrete lining, i.e., a top-down panel lining and a concrete, monolithic, bottom-up lining using panel forms. The concrete lining was laid on a 2.6-m-thick base ring beam set between depths of 632.4 and 635.0 m (**Figure 1**).

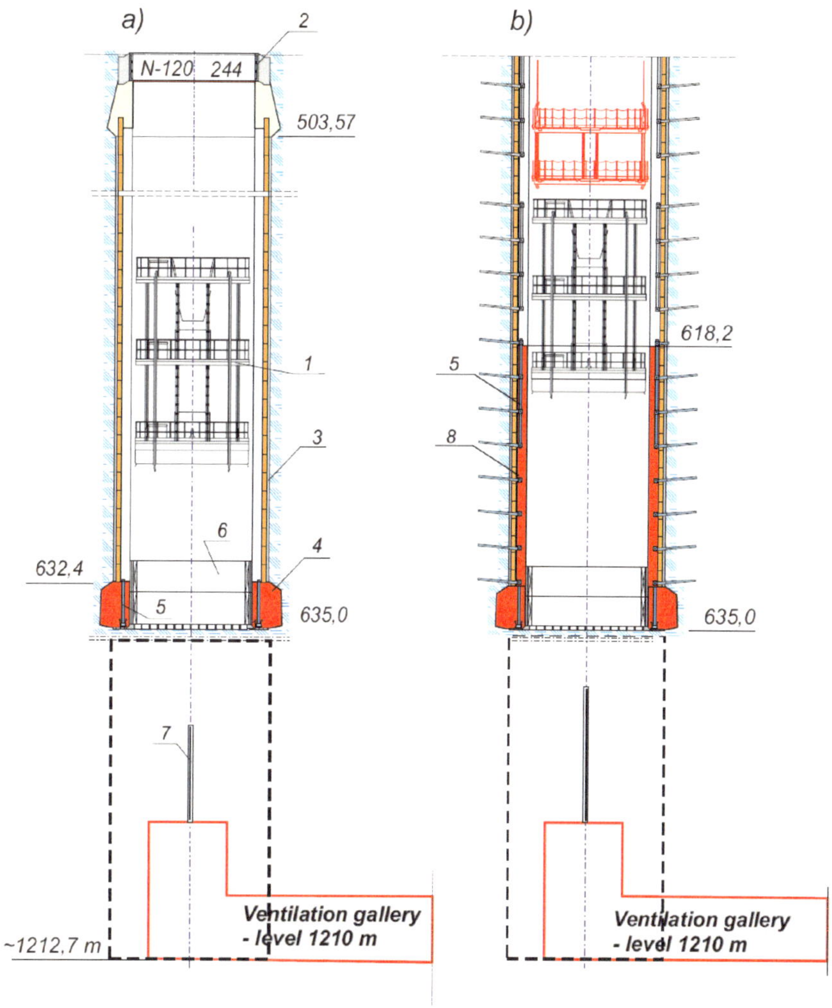

Figure 1.
The last phases of shaft sinking in the frozen rock area. (a) Installation of foundation for the final shaft concrete lining. (b) Section of final shaft concrete lining with drainage. Explanations: 1, three-deck shaft working platform; 2, cast-iron shaft lining; 3, preliminary pre-cast segmental shaft lining; 4, shaft lining foundation; 5, boreholes in the drainage system; 6, sliding formwork H = 3.75 m; 7, dewatering borehole TS-1 (d = 3.5," L = 576 m); 8, final concrete lining.

According to the records [6, 7], the in situ rock temperature at a depth of 632 m was about 32°C. So, it was reasonable to expect that the end portion of the frozen mantle would also be exposed to increased heat from below. When the shaft face reached a depth of 632 m without any difficulties, it seemed that the most challenging section had been sunk as designed and on schedule. Yet, nature retained its unpredictability.

After the two last reinforced concrete panel rings had been completed, with excess material excavated to make a curb gap 4 (**Figure 1**) at a depth of 632 m, a small water leak, estimated at about 3–5 L/min, was noticed at the shaft bottom at the thill sidewall interface. The water was clean, very cold, and slightly salty. For a shaft sunk using the freezing method, in which the freezing core usually has a temperature below −15°C, this was unusual and perplexing. Since the freezing pipes had reached a depth of 635 m, no liquid water should have occurred at a depth of 632 m. However, this phenomenon could be partly explained by the water's salinity. Unfortunately, the electrical conductivity of this water has not been documented. In these circumstances, the TS-1 dewatering borehole work was intensified. Also, work commenced on the final concrete lining 8 (**Figure 1**)—constructed from the bottom up—equipped with a drainage system [7].

It was found that even though all the freezing safety requirements had been observed, the ice mantle along this section was not completely watertight and did not fully prevent water inflow into the shaft face. The movement of slightly saline water at a temperature above zero ($t_w > 0°C$) caused the frozen mantle to be soaked from below and consistently thawed, with water inflows effectively increasing day by day. The situation was becoming dangerous, as no shaft pipe drainage had been planned down to this depth. This meant that the shaft had no pipelines through which the water could be pumped up to the surface. The further section of the shaft was designed to allow drainage via the TS-1 dewatering borehole drilled from a level of 1212.7 m (**Figure 2**).

The increasing inflow of water was diverted to the surface using only buckets. After about 2 weeks of shaft work involving the construction of a concrete curb at a depth of 635 m and the construction of an 18 m final concrete lining, water inflow into the shaft had increased to about 700 L/min. In this situation, it was impossible to continue any work in the shaft other than intensive dewatering using of buckets. Ultimately, this measure did not save the shaft from partial flooding. The water table in the shaft stabilized at a depth of 533.0 m, which means that the water column was 102 m (see **Figure 2**).

Due to the prolonged length of the 564 m TS-1 dewatering borehole and the water level reaching 533 m (**Figure 2a**), the decision was made to use a high-performance RITZ submersible pump (HDM 6723/11DPF). Installed 4 weeks later, with a capacity of 15 m^3/min, the submersible pump succeeded in quickly dewatering the flooded shaft section (**Figure 2b**). Also, after 2 months of further work, the water inflow into the shaft was found to have reached 2.5 m^3/min. The dewatering borehole TS-1 (**Figure 2**) was successfully completed almost at the same time the shaft was dewatered using the submersible pump. After 6 weeks of intensive and highly precise drilling work, the borehole reached the shaft bottom, located only 0.5 m from the shaft axis. By this point, the water inflow had increased to 3.0 m^3/min. Since the water inflow was expected to increase further, the decision was made to drill a second dewatering borehole—TS-2 (**Figure 2c**). Due to the considerable water hazard associated with a water inflow of 3.0 m^3/min, it was also decided that the section with a waterproof tubing lining be extended to the 650 m level. In addition, the decision was made to comprehensively grout the entire area affected by the substantial water inflow.

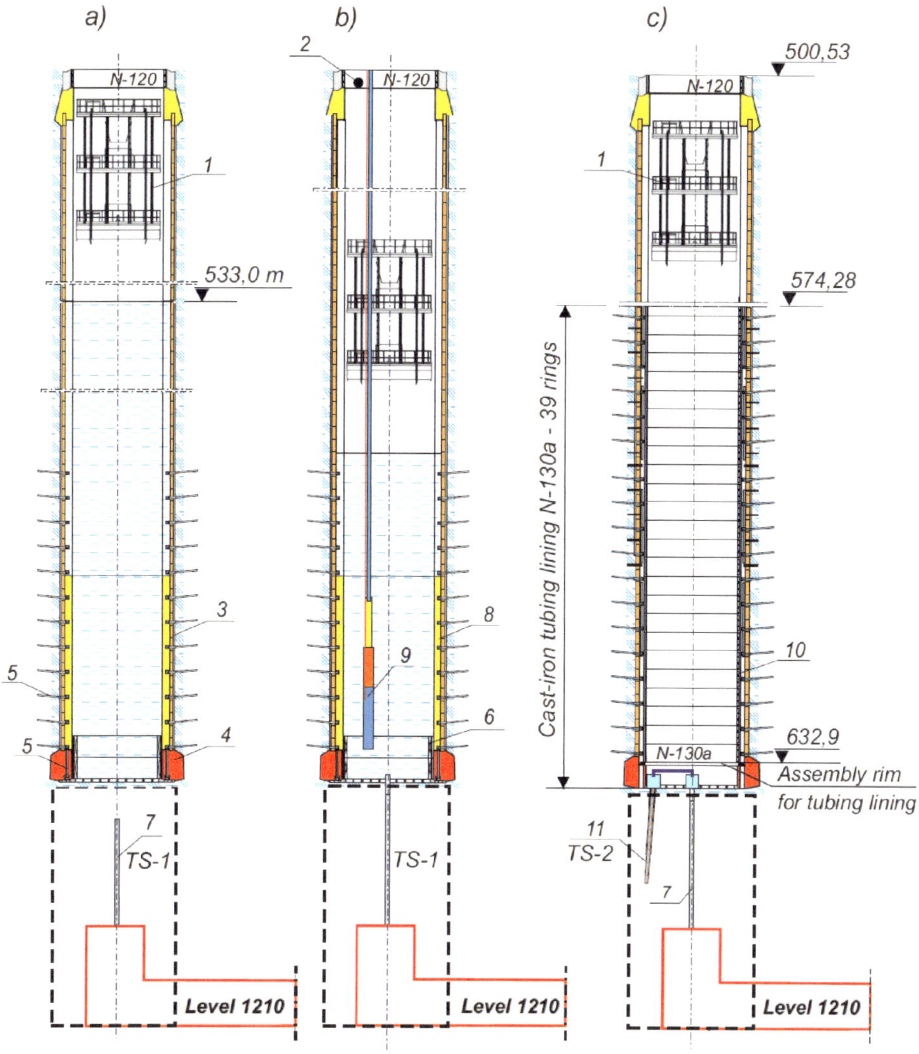

Figure 2.
The phases of dewatering the sunken shaft section. (a) Shaft flooding. (b) Dewatering of the shaft using a "RITZ" submersible pump, (c) replacing the concrete lining along the 574.3–632.9 m section with a cast-iron tubing lining. Explanations: 1–8, see ***Figure 1****; 9, RITZ submersible pump; 10, cast-iron tubing lining; 11, TS-2 dewatering borehole.*

2.2 Removing the causes of the water inflow

The substantial water inflow forced the shaft construction company to both further redesign the shaft lining and adjust the sinking technology along the 635–650-m-deep section. Apart from the costly dewatering, one of the direct effects of the partial shaft flooding was the need to redesign the lining in the flooding area (**Figure 2**). The concrete panel lining was replaced by a tubing panel lining, with a concrete tube set between them (**Figure 2c**) [7]. Due to this replacement, it was additionally necessary to:

a. Demolish the completed 18 m section of the concrete lining above the curb, at a depth of 635 m, without damaging the preliminary panel lining.

b. Partially demolish the curb at a depth of 635 m and mount a steel ring beam on the curb's foundations to lay the first tubing ring.

c. Construct the lining of 39 N-130a tubing rings from the bottom up, to a depth of 574.3 m, without damaging the preliminary panel lining, the initial step being to lay the first ring on the steel ring beam in the curb at a depth of 635 m.

d. Complete the grouting work above the curb, at a depth of 635 m, so that the shaft could be safely sunk along the 635–650 m interval.

e. Sink the shaft along the 635–650 m tubing-lined section, including constructing a curb at a depth of 650 m.

f. Construct the lining of N-120 cast-iron tubing rings from the 574.3 m level upwards to the point of connection between the picotage gap and the upper tubing column at a depth of 500.47 m.

In the first phase, the rock behind the lining was grouted using multiple techniques. In the first phase, 3-m-long holes were drilled in rings 309 and 310 through cement plugs in the tubing lining. A total of 26 t of cement grout were injected behind the lining through these holes to separate the upper water horizons from the problem area of the shaft.

In the second phase, the cement grout was injected behind the lining along the 617.9–635.0 m section, using 2-m-long horizontal holes drilled through the concrete plugs, 10-m-long horizontal holes drilled through the cement plugs, and 15-m-long inclined holes drilled at an angle of 40° through the concrete plugs. Due to the very substantial water inflow from this area, "Ekopur HW" quickset two-component

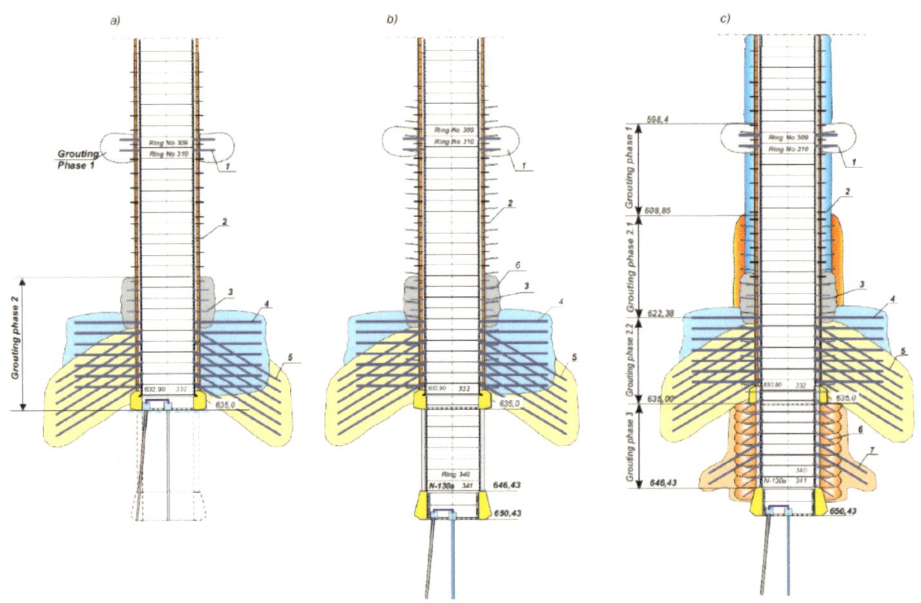

Figure 3.
Grouting process and shaft sinking along the 635–650 m section. (a) Grouting along the 598.4–635.0 m section, phase 1 and 2; (b) shaft sinking along the 635–650 m section, (c) grouting along the 635–650 m section, phase 3. Explanations: 1, cementation of the rock behind the lining (insulating layer) in N-130a tubing rings 309 and 310; 2, grouting of the rock behind the tubing lining through concrete plug holes in the tubing; 3, grouting of the rock and tubing lining through "cementation" holes in tubings (2-m-long horizontal holes); 4, grouting of the rock and tubing lining through "cementation" holes in tubings (10-m-long horizontal holes); 5, grouting of the rock and tubing lining through concrete plug holes (15-m-long inclined holes); 6, grouting of the rock and tubing lining through "cementation" holes in tubings (2.0-m-long inclined holes), 7, grouting of the rock and tubing lining through "cementation" holes in tubings (10.0-m-long inclined holes).

Phase	Grouting materials used [Mg]		
	Cement	EKOPUR HW polyurethane	Total
Phase 1	45.5	24.5	70.0
Phase 2	33.3	14.8	48.1
Phase 3	16.4	0.32	16.7
Phase 4	342.7	32.4	375.1
Total	**438.0**	72.1	**510.1**

Table 2.
Grouting materials used to prevent water inflow [7].

polyurethane adhesive was used in addition to the cement grout. The grouting work is illustrated in **Figure 3** [7]. Once the 635–650 m section of the shaft had been sunk, a curb was made in the tubing lining, comprising 130a tubings (9 rings) installed from the top down at a depth of 650 m, and a shaft face dewatering system was installed using boreholes TS-1 and TS-2 (**Figure 2**).

In the third phase, cement grout was injected behind the lining along the 635–635.0 m section (**Table 2**). Then, as part of the fourth grouting phase, the entire 574–500 m section of the tubing lining was sealed. It took a total of more than 500 t of materials (**Table 2**) to complete the grouting process.

3. Eliminating water hazards associated with the S-1.3 shaft sinking project in the Lublin Coal Basin

The hydrogeology of the Lublin Coal Basin is highly complex, and mining in this area is challenging. At 710 m from the surface, the coal measures are covered by heavily waterlogged Jurassic, Cretaceous, and Quaternary formations. This has considerable implications for mine shaft sinking. A simplified geological profile is presented in **Figure 4**.

All the shafts in this basin had to be sunk using the freezing method, at least along the 0–180 m section [4, 6, 8]. The first shafts for the Bogdanka Mine were given the numbers S-1.1, S-1.2, and S-1.3. After the S-1.1 shaft had been sunk to a depth of 960 m, a disastrous water leakage occurred from the connector pipes left in the lining, which caused extensive flooding. Consequently, it was necessary to fill in and abandon that shaft. Drawing on the S-1.1 experience, the S-1.2 shaft was sunk to the target depth of 995 m without any major difficulties. Although the flooding of the S-1.1 shaft had also caused partial flooding of the S-1.2 shaft through the galleries already sunk to a depth of 960 m, the dewatering proved to be fairly easy. The sinking of the S-1.3 shaft might be the most interesting and perhaps the only such case in the global history of shaft construction, as it ultimately required simultaneous rock freezing in the lower section and rock heating in the upper section. Below is a detailed discussion of how this was done.

3.1 The S-1.3 shaft sinking

The experience gained sinking the S-1.1 and S-1.2 shafts indicated that it was possible to use a different technology, more based on the traditional sinking method, which is much less costly. A decision was made to freeze the rocks along the 0–180 m section before constructing the first section, as it passed through the Quaternary strata and the highly waterlogged layers of Cretaceous formations, with a maximum

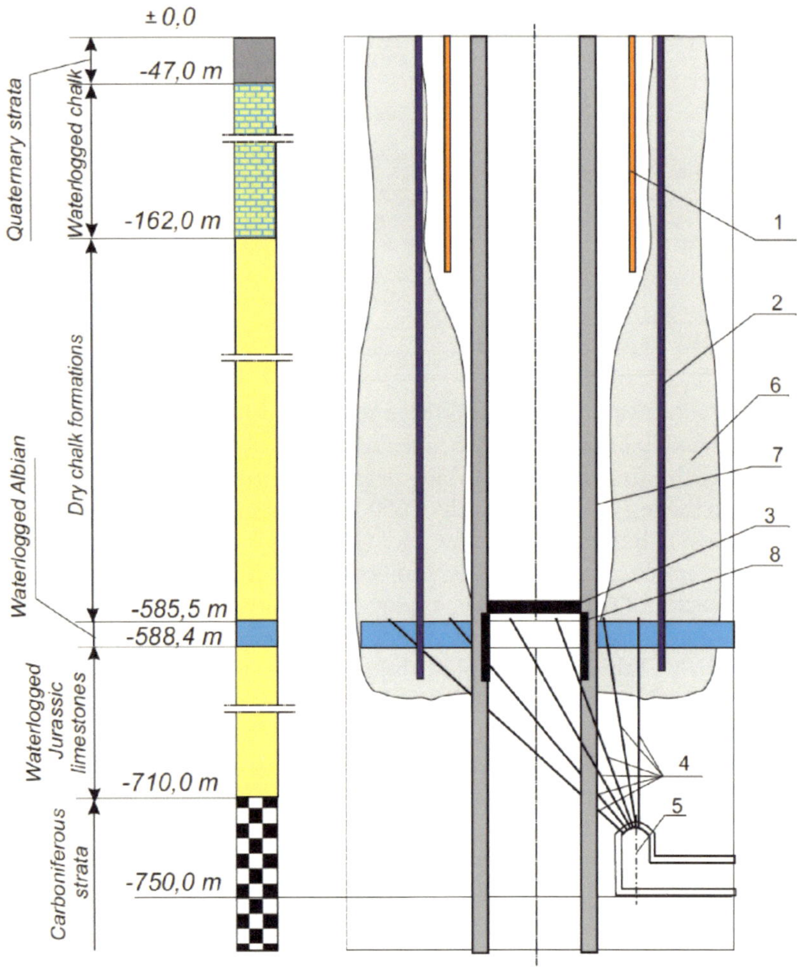

Figure 4.
Diagram of the S-1.3 shaft sinking using both rock freezing and rock heating. 1, boreholes for rock heating along the 0–180 m section; 2, boreholes for deep freezing along the 0–570 m section; 3, concrete plug above the Albian layer; 4, drainage boreholes in the Albian layer; 5, working with a drilling chamber at a depth of 754 m; 6, frozen rock mantle; 7, final concrete panel lining; 8, cast-iron tubing lining along the Albian formation section.

depth of 162 m (see **Figure 4**). Below the 180 m level, the plan was to sink the shaft to a depth of 570 m using conventional method, i.e., without rock freezing. This would substantially reduce costs. The biggest puzzle, and, as the construction company would see, the greatest challenge involved in this shaft sinking project was the thin (≈2.9 m) Albian layer (**Figure 4**), which was composed of sandy-lime quicksand with a water pressure of about 5.5 MPa. An assumption was made that a shaft working could pass through such a thin layer of waterlogged formation once the layer had been provided with borehole drainage system (**Figure 4**) drilled in the working at a depth of 754 m. With the drainage system in place, the pressure could be reduced, making it possible to petrify both the Albian formations and the Jurassic formations deposited underneath, all the way to the Carboniferous roof. This way, the shaft could be sunk conventionally down to the target depth of 1035.45 m.

Here, we should warn those who are enthusiastic about using grouting, regardless of the conditions. In this specific case, the company constructing the shaft failed to provide the mentioned formations with a drainage system. In effect, it became impossible to chemically petrify the Jurassic formations any further, and the only viable sinking option left was the freezing method. At this

Parameter	Design	Actual
Freezing time, months	9.2	About 7
Amount of energy consumed to create the frozen mantle, MJ	71,310,951	45,638,000
Heat supplied through the boreholes along the 14 m diameter circle (0–180 m), MJ	15,323,000	12,509,800
Heat supplied through the air supply duct to the warm-air shaft, MJ	23,197,000	16,130,000
Total energy consumed, MJ	109,830,951	74,277,800

Table 3.
Projected and actual energy consumption in the process of rock freezing when sinking the S-1.3 shaft [3].

point, the expected substantial savings stemming from the use of a different shaft sinking method were no longer viable. In addition, the construction company faced the problem of refreezing within the 0–570 m zone, where the final lining had already been laid. The Polish engineers involved in the project knew that the refreezing of rocks would produce a great pressure surge on the lining, effectively destroying it [6].

The engineers considered it necessary to drill 43 additional boreholes at a depth of 610 m. These had an unusual diameter of 308 mm and were drilled in an 18 m diameter circle [3, 8]. Also, an unprecedented decision was made to use sectional freezing—an approach which, although theoretically known and viable, had not been applied in shaft construction before. Thus, the boreholes were fitted with two freezing pipe columns and a column of downcomer tubes inside a 139.7 mm diameter column. They were properly sealed so that the brine could circulate only in the lower parts of the boreholes, below 570 m.

Regrettably, this plan failed, too. The shrinkage stress in the steel due to the low temperature of the brine caused the outer column to leak, allowing water to enter the borehole. This complicated the whole process of section freezing, making it necessary to reconsider freezing along the entire depth of the shaft. As feared, the freezing caused damage to the lining along the 0–570 m section soon after commenced. At this point, the decision was made to apply a globally unprecedented solution, in which the lower section of the shaft was frozen, while the upper part of the shaft, along the 0–180 m section, was heated with warm water. To provide the inflow of warm water, the engineers used the boreholes drilled to freeze the first section of the shaft along a circle with a diameter of 14 m. The work diagram is presented in **Figure 4**.

Ultimately, this unprecedented project proved a technological success. However, although the shaft was eventually sunk, the project can hardly be described as successful, given the completion period of almost 10 years and the substantial energy costs involved. The substantial costs of sinking the S-1.3 shaft are reflected in the amount of energy consumed in the process of rock freezing and heating. These parameters are shown in **Table 3**. It should be noted that the actual values were much lower (by about 38%) than the design values.

4. Conclusions

This paper shows how changeable and unpredictable hydrogeology can lead to challenging and very costly problems in shaft sinking projects. In the case of the polish shafts, a water hazard that had not been accurately identified by hydrogeological surveys led to a number of adverse effects. These included the substantial amount of grouting materials used, the extended project completion period (it took almost an additional year to finish the project), and the need to replace the concrete panel lining with tubing lining along a 150-m-long section of the shaft.

This paper presents case histories that should serve as the ultimate warning against underestimating the projected inflow of water into a shaft during its sinking. A number of shaft construction projects recently implemented in Poland further illustrate this point. Since water inflow projections proved inaccurate, it is necessary to improve the accuracy of hydrogeological surveys in the areas where mining is planned. Poland has extensive and highly informative experience in successfully dealing with water hazards related to shaft construction.

Author details

Piotr Czaja[1], Paweł Kamiński[1]* and Artur Dyczko[2]

1 AGH University of Science and Technology, Kraków, Poland

2 Mineral and Energy Economy Research Institute, Polish Academy of Science, Kraków, Poland

*Address all correspondence to: pkamin@agh.edu.pl

References

[1] Ciania Z, Lekan W, Wójcik J. Planned and interventional cementation treatments in shafts sunk within the Lublin Coal Basin. In: Proc Symp: Experiences in the Use of the Grouting Technology in Rock Sealing, Reinforcing and Petrification in Underground Construction and Workings. Katowice: SITG; 1988

[2] Czaja P, Kohutek Z, Wichur A. The water-related problems and difficulties encountered when exposing deposits – Coping methods. In: The Hydrogeology of Polish Mine Deposits and the Water-Related Problems in Mining. Kraków: Uczelniane Wydawnictwa Naukowo-Dydaktyczne AGH; 2003. pp. 146-200

[3] Kohutek Z, Wichur A, Wilk Z. Water-related issues when exposing deposits. In: Wilk Z, editor. The Hydrogeology of Polish Mine Deposits and the Water-Related Problems in Mining. Kraków: Uczelniane Wydawnictwa Naukowo-Dydaktyczne AGH; 2003. pp. 424-435

[4] Krywult J, Wichur A. A simplified method for designing preliminary linings for shafts sunk using rock freezing. Przegląd Górniczy; 1993. p. 5

[5] Stachowiak-Maciejowska K, Rożek R. The R-XI shaft is the 29th shaft of this Polish copper mining company, sunk by PeBeKa S.A. Nowoczesne Budownictwo Inżynieryjne. 2005;**09**:35-36. Available from: http://www.nbi.com.pl/assets/NBI-pdf/2005/2/pdf/9_pebeka.pdf

[6] Wichur A, Czaja P, Poprawski W. Guidelines for designing the frozen mantle thickness. OBRBG Budokop w Mysłowicach, Konferencja naukowo-techniczna Budownictwo górnicze i podziemne w nowych warunkach gospodarowania, Materiały konferencyjne i referaty problemowe, Kokotek k/Lublińca 16-17. IX. 1991. pp. 32-38

[7] Kosmalski M, Kulicki J, Stróżyński M. The elimination of increased water inflows into the R-XI shaft in the Zakłady Górnicze Rudna mine. Geoinżynieria i Tunelowanie. 2005;**1**:46-54

[8] Kicki J, Dyczko A. 30 years of LW "Bogdanka" S.A. In: The History and the Future. Krakow: Bogdanka;2012

Chapter 7

Anchorage Pile Strengthening of Shale Slopes and Cementing Falling Stone Blocks by Mixture of Melted Waste Plastics/Asphalt and Fly Ash for Slope Stability in Asphaltite Open Pit Mining Site in Avgamasya, Şırnak

Yildırım İsmail Tosun

Abstract

The Rock Fallings, Shale Slopes Stability, and Stability Risk Assessment in Şırnak open pit asphaltite mining should be searched in detail and improved in several coal mining sites in Şırnak Province, reaching over 120 m height with 60–65 degree shale slopes, developing major landslide in the open pit Şırnak open pit coal mining. The rock fallings endangered the mining safety in recent years. This research provided stability patterns and cementing method strengthening cracks. The stages of experimentation explored the geo-technical characteristics and geoological formation. Fort his aim, four different open pit mining areas with similar geotechnical conditions, two main strengthening methods, and patterns of researches were developed. Firstly, data on landslides and rock dynamics over explosions were followed, and secondly, as happened commonly in the past, the same geological, geomorphologic, hydrological, climatic conditions were taken. Anchorage pile strengthenıng of slopes and cementing falling stone blocks were performed by mixtures of melted waste plastics/asphalt and fly ash for stability of higher slopes over 120 m height and over 65 degree in asphaltite mining site in Silopi and Avgamasya open pit No.1 mining site in Şırnak were carried out. On the other hand, due to that creep style rock falling from top of slopes, those melted polymer cementing of anchorage bolting and cracks, to eliminate those falling failure types and features, will be advantageous. The unconditional expectations related to this study was also defined for this region, such as the influence of the ground water, rock cracks and slope design, and explosion exchange dynamics leading to landslide. GEO5 software and manual stability analysis showed high risk area for plotting.

Keywords: mining pit, Şırnak asphaltite, active potential landslide, mining, geotechnical stability, slope stability

1. Introduction

Engineering geotechnical properties of surface units were determined by making geological map of Şırnak province and surrounding areas [1–5]. Geological mapping of slopes on large-scale topographic maps was one of the first landslide studies in the region. By determining the engineering properties of the slopes, it was aimed to draw attention to the importance of the area and the geotechnical criterion in the construction of the municipal development plans in the future.

Landslide is the downward slope due to the effects of massive soil masses or rocks on sloping slopes in mountainous areas such as gravity, slope, water, climatic factors, tectonics, weathering [6–9]. The geology of the material can be listed as precipitation, erosion, earthquake, and vegetation deprivation. Limit stress and balance analyses give accurate results in determining the landslide hazard and predicting future landslides [10].

In areas with high danger, landslide and related events will increase proportionally with increasing population density. It is very difficult to eliminate and reduce the risks arising from the processes of these landslides. There is an urgent need to better understand the character of the operation safety in open pit coal mining and to develop more predictive tools for stability.

Processes such as heavy rains, seismic, changes in groundwater level, erosion, climate, weathering, and natural topography are the natural parameters that trigger landslides. These effects increase the shear stress or decrease the shear resistance of the slope material [11]. Another important parameter that triggers landslide is urban activities. Increasing population and creating new living spaces forced people to settle on the slopes that present geological danger. The realization of equipment uses such as installing, practice, the creation of safe areas, and the realization of stress structure activities brought on by the excavation developed on the slopes can disrupt stability and create human activities and explosions that trigger landslides.

Especially in developing countries, the land in mountainous areas is not used in accordance with the topology, and wrong land use increases the probability of landslide development. Sustainability cannot be achieved in terms of physical environment, change and efficiency of landslide risk areas in open pit mining sites, and operation safety.

Important landslides developed in open pit coal mining of the country in recent years are determined based on ground conditions. The stability of working area was managed with different methods such as geotechnical characters and geological formations, precautionary measurements, and the processes determined. Researchers have two basic theories for areas with similar geotechnical conditions [12–30]. Firstly, landslides are formed in the same geological, geomorphologic, hydrogeological, climatic conditions as in the past. Regarding the past phenomenon, the stability studies were carried out. Another is that the types and properties of landslides will be the same with other open pits. Therefore, knowing the mechanism and properties of past landslides is important basic information to evaluate landslides that may develop in the future, neighboring regions, or geotechnical similar areas. Geological and geotechnical analyzes of the slopes should be carried out in order to minimize the economic and social losses and casualties caused by landslides. In this direction, within the coal mining area of Şırnak Avgamasya and Silopi, open pit mining was carrying high landslide or rock falling risk (**Figure 1a**).

The geotechnical properties of the slopes where landslides occur in the districts that are 0.2–0.4 km from the south of the city and the center are analyzed, and the stability analyses were carried out with different methods using the GEO5 program. Within the scope of this project, a 1/5.000 scale engineering geology map covering 0.07 km^2 of the study area and its surroundings that will be opened to urban use has been prepared as a

DOI: http://dx.doi.org/10.5772/intechopen.91987

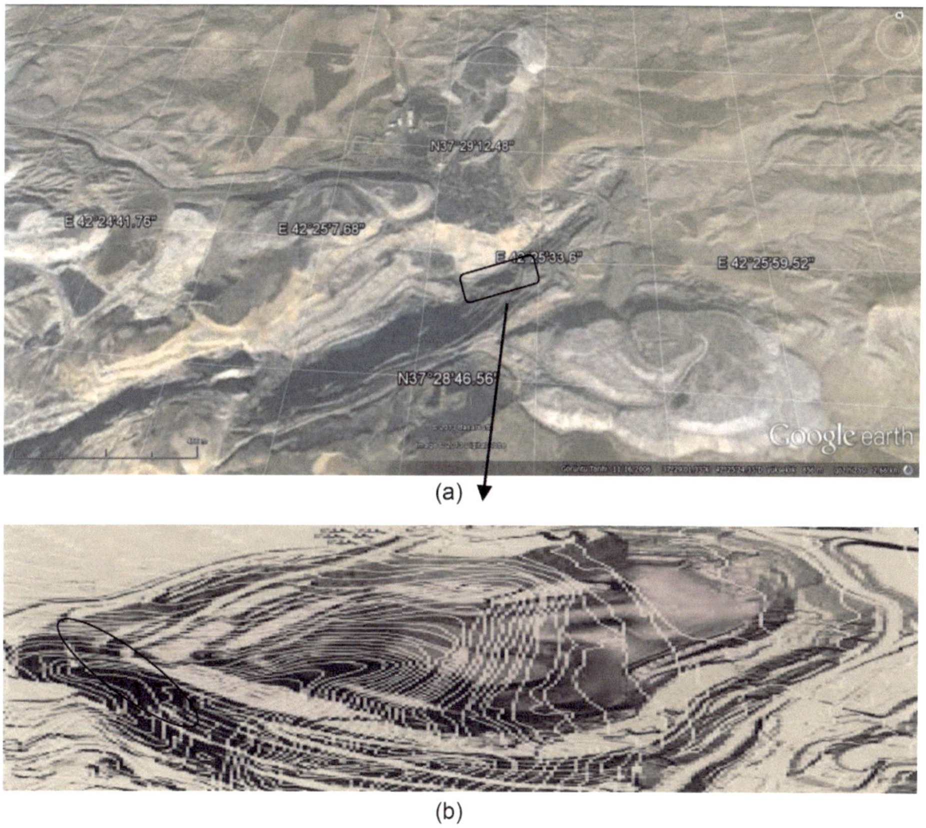

Figure 1.
(a and b) Satellite and contour topography of Avgamasya No.1 Pit Şırnak asphaltite coal mine site and survey area 1/18000 and 1/5000.

result of field and laboratory studies, and a topographic map has been created for all four hills by using the polar coordinate system. The high risk of rock sliding or falling damaging the asphaltite coal production occurred at the local open mining site 1 in Avgamasya seen as shown in **Figure 1b**. Black area was coal extracting area.

2. The coal geology in Avgamasya, Şırnak

Due to its tectonic structure and stratigraphy in the Southeastern Anatolia region, besides having reservoir rock and cover rock properties required for hydrological conditions, the water going down to the depths of the ground along the stretch cracks in the region are in a position to provide the necessary fluid for open pit asphaltite mining [31]. The crust of the earth was subjected to a stretching in the east-west direction with the effect of compression in the north-south direction throughout the region, and olivine basaltic magma rose from the asthenosphere along the stretch cracks formed.

Basaltic magma, which reaches to the surface in the Karacadağ region between Diyarbakır-Şanlıurfa-Mardin, Gaziantep Yavuzeli region, and İdil-Cizre regions, flowed in several phases and left large areas under lava flows. Magma, which does not reach the earth, as in the north of Batman, created hot areas by making intrusions in several places. This situation in Şırnak is not clearly seen in the geological map of the region made by MTA General Directorate. The four lithostratigraphic units in the study area were distinguished from late period time age as Mardin

Vulcanite (Upper Miocene) [30], Old Alluvium (Quaternary), New Alluvium (Quaternary), and Slope Rubble (Quaternary). Consisting of vulcanite, tuff, agglomerate and andesitic basaltic lava, syenitic rocks, which make up a large part of the study area, it shows as much shale and porous rock formations and slopes as occurred in the studied area chosen.

In recent years, both the opening of our university in our city and the migration movement from rural to provinces have also affected the coal economy in Şırnak. Open pit mining excavation and asphaltite production increased in different excavated pits in the area. A total of 500,000 asphaltite excavation by Asphaltite open pit mining with 20 separate pits continued to increase rapidly in Şırnak. Equipment and safety demands have also increased due to excessive extraction. This increase in demand caused people who are not competent in excavating in open pit mining production to enter the hauling, the control mechanism of dumping ability, modeling slopes to control the intensive stability, and areas that are not included in the development plan that are rapidly foreseen to mining development and safety. A vast majority of the excavatings in the Avgamasya open pit No. 1, 2, 3, 4 were excavated without adequate ground research. In the asphaltite mining of Şırnak, there were generally adjacent open pits in Silopi, Uludere. It is believed that such new open pits in Şırnak have been developed due to the neighboring hard condition of rock cracks, rock fallings, landslides, and sometimes the existing collapsed pits (or structural damage caused by the rotation of the equipment) or hydrological ground problems. Finally, Şırnak Avgamasya No. 2 landslide caused 8 workers' death, closing the excavation work. For these reasons, the investigation of the new buildings to be built in our province and the ground conditions on the basis of regional and parcels of new areas to be developed has become essential [18, 19]. The unconditional expectations related to this study were also defined for this region such as the influence of the ground water, rock cracks and slope design, and explosion exchange dynamics leading to landslide. GEO5 software and manual stability analysis showed high risk area for plotting (**Figure 2**).

Another issue for the province of Şırnak is the causes of the damages that may occur in possible explosions and earthquake shock waves caused by human conditions which are due to the engineering errors as well as the lack of control mechanism and construction defects in the excavating process. Şırnak Avgamasya open pit No. 1 mining site is located in the 2nd degree earthquake zone and these shortcomings mentioned above are also seen in our city. As seen in **Figure 3**, the risk of slope stability risk in explosions and excavation near earthquake district of 1

Figure 2.
Avgamasya No. 1 pit Şırnak asphaltite coal mine site and survey area.

Figure 3.
North face of Avgamasya No. 1 pit Şırnak asphaltite coal mine site and survey area.

degree in Uludere Ortabağ districts. Cizre and Merkez were areas of 2 degree risk. The studies considered to take precautions for steep slopes avoid those as below:

- steep slope structures over 60°;
- earthquake resistant hollow-type structures;
- high slope type pits, excavating without basement control in real sense; and
- slope models designed without paying attention to hydrologic conditions and ground conditions did not escape attention.

Ground movements that may occur on a regional and/or 5–10 m basis have been observed in many irregular shale facing, high crack risk areas as seen in **Figure 4**.

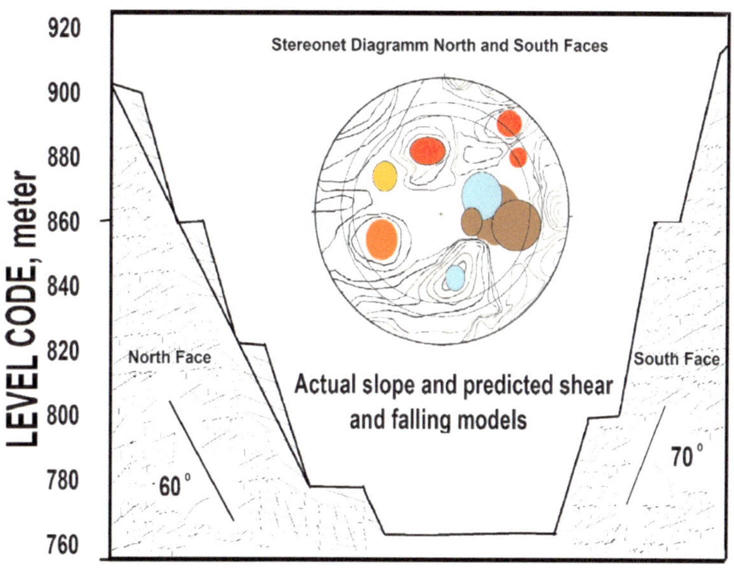

Figure 4.
North and south steep slope faces of Avgamasya No. 1 pit of Şırnak asphaltite coal mine site and survey area.

3. Geotechnical properties of slopes and modeling

3.1 Field studies

It is observed that the city center of Şırnak is located on a topography inclined to the south. Formations generally are composed of claystone and siltstone in the field.

It was observed and it is known that the center of Sirnak province is in Germav Formation. Germav Formation causes rapid formation of landslides as it forms a high slope topography by rapidly eroding due to its wear resistance. Therefore, in summary, Şırnak city center is located on the anchored Germav Formation, which is a mixture of sandy, carbonated, clayey, and silty units due to old landslides. Slope debris extends from the south of the stream (**Figure 2**) to the boundary of the working area. It has been determined with the field observations that the slope debris is composed of myocene limestones. Their thickness is extremely variable. They outcrop relatively less in areas where slope inclination decreases.

The surface of new alluvial deposits starting from city center to the south of Avgamasya open pit 1 (**Figure 3**) in the study area is gray marl shale. This part is generally covered with silty soil, and some parts of it consist of sandy and clayey zones. It has been determined that the new alluvium continues in the drilling up to 35 m by the Special Provincial Administration [32]. Slope Rubble is located south of the Stream in the study area.

Grain sizes range from fine clay to coarse sand. The thickness of the rubble, which does not show any grading and grading, varies between 10 and 35 cm. The active and potential landslide areas observed in the slope debris have been studied in detail.

3.2 Geotechnical properties

American Standards (ASTM 3080) were taken as a basis in the experiments carried out to determine the geotechnical properties of the surfaces surfaced in landslide risk areas. In the landslide risk areas of the study, undisturbed and lump rock samples were taken from the ground parts of the slope face. In experiments on the lines shown in **Figure 5**, the slope unit weight in wedge style and circle shape were considered in analysis with manual and GEO5 software and safety consistency limits were obtained.

The slope model construction plan is shown in **Figure 6**. The anchorage improved stability safety factors for steep sliding slopes in 20 m height excessive to 30 m. Wire mesh hanged top of slopes were avoiding rock falling of highly cracked shale block stones at 3–5 m size. The pile anchorage was designed and practiced for

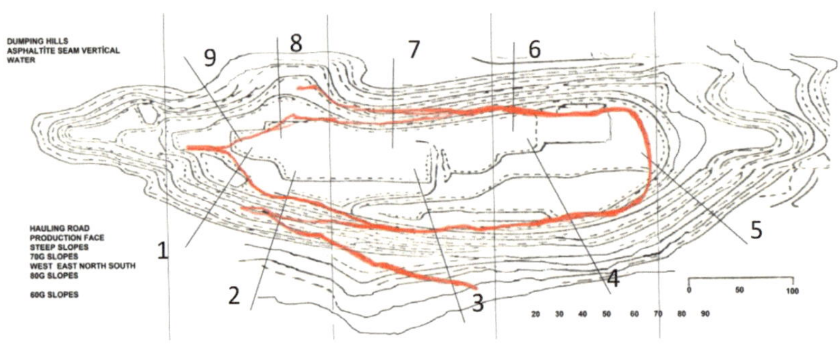

Figure 5.
North and south steep slope faces of Avgamasya No. 1 pit of Şırnak asphaltite coal mine site and survey area with anchorage pattern.

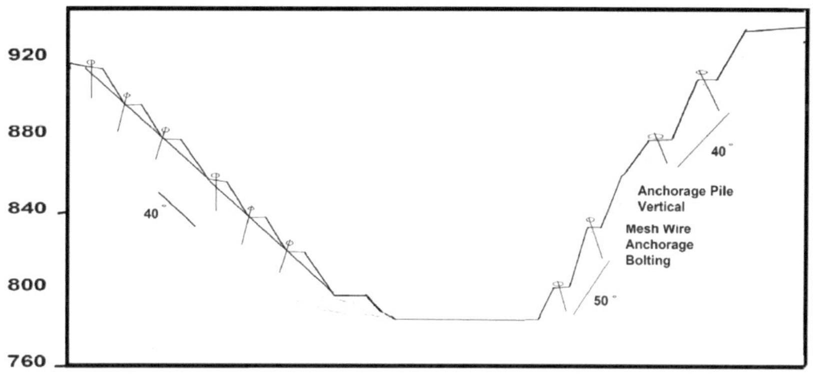

Figure 6.
North and south steep slope faces of Avgamasya No. 1 pit of Şırnak asphaltite coal mine site and survey area with anchorage pattern.

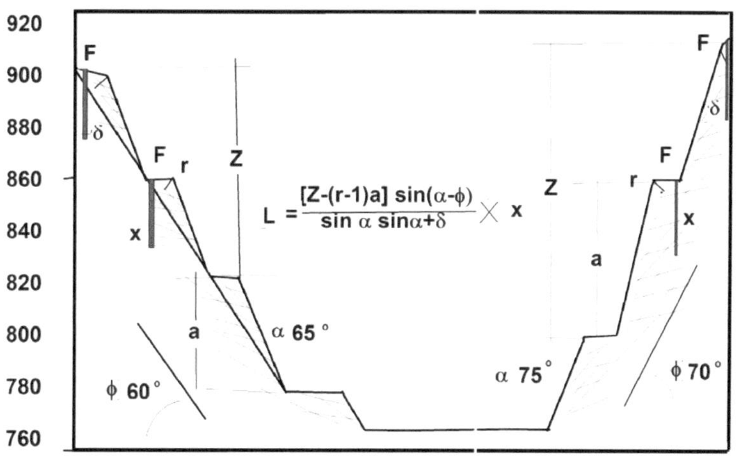

Figure 7.
North and south steep slope faces of Avgamasya No. 1 pit of Şırnak asphaltite coal mine site and survey area with anchorage pile pattern.

hauling road slopes control and stability at the constructed near deep of slopes bottom line as shown in **Figure 7**.

The uniaxial compressive test point load tests, RQD values, and Mohr-Coulomb criteria diagrammed were determined regarding ASTM standards; effective cohesion (c‘) and effective shear resistance friction angle (ϕ°)’ of the rock samples were given in **Tables 1** and 2 on the standard rock samples of 50 mm cubic forms with the help of ELE press equipment [23–25].

Undisturbed rock samples used in the point load experiment were classified according to ASTM standards and determined as cracked, altered ground. Rock samples show nonplastic character and there are also coarse-sized gravel.

In addition, during the making of these experiments, the unit volume weight, the amount of compression, and the void ratio were determined.

The results obtained in plastic and liquid limit tests are given in **Table 1** for each sample. According to the ground classification in North district, the slope samples in the landslide N1 are in the less plastic shale and not plastic group, whereas in the region containing the landslide hazard N2 and N3, it is determined that there is less plastic.

Rock formations	Thickness (m)	RQD (%)	c′ (kPa)	φ′	P1 (MPa)	I1 (MPa; 50 mm)	Shear strength (mm/s)	γsat n (g/cm^3)	γdry (g/cm^3)
S1	25	25.9	3700	22	22.0	1.4	22	2.92	2.58
S2	74	42.9	3300	28	15.0	1.8	13	2.92	2.57
S3	25	40.8	2300	32	26.0	1.7	14	2.9	2.62
N1	47	25.9	2700	18	38.0	1.0	25	2.92	2.61
N2	55	35.4	4700	17	33.0	2.3	12	2.8	2.68
N3	46	33.9	4100	14	36.0	2.2	12	2.8	2.68

Table 1.
Results from geotechnical tests on samples taken from landslide slopes.

Rock no	S1	S2	S3	N1	N2	N3
γ_{sat} max (g/cm^3)	2.98	2.75	2.77	2.98	2.75	2,77
w_{opt} (%)	15.9	11.9	11.0	12.3	3.8	3,3
Permeability (k) (cm/s)	5.3*10–4	3.0*10–5	6*10–5	1.3*10–4	3.3*10–6	5.2*10–6

Table 2.
Proctor of ground samples and permeability test results.

The results obtained from the geotechnical tests carried out on the samples taken from the slopes forming the landslide threat are given in **Tables 1** and 2.

4. Results and discussions

The water content on the ground will be significantly affected by the clay content. When evaluated according to the percentage of clay in the floor, the floor samples show a non-cohesion or low cohesion feature.

The grain volume weights obtained by the experiments done on the samples taken from the landslide sites are shown in **Table 1**. In order to determine the soil types according to the grain size, grain distribution experiment has been carried out and the results and names in the combined soil classification are given in **Table 1**.

In order to determine the permeability of the ground, a fixed level permeability test instrument was used. The degree of permeability of the ground was determined by evaluating the results of the experiment (**Table 2**). When **Table 2** is examined, it is seen that the slopes S1, S2, and S3 fall under the permeable ground class.

The γ_k and w_{opt} values obtained as a result of Proctor experiments on soil samples taken from landslide areas are given in **Table 2**. With this experiment, optimum water content on the ground and maximum dry unit volume weight are determined and used for stability calculations of the slopes. Compaction parameters do not affect the stability of a natural slope, because these parameters are the parameters of the ground compacted in the desired way. In artificial slopes, compression parameters are used directly. If there is a landslide danger in a natural slope, in case of compression, the stabilization analyses are compared using these parameters. In the precautions to be taken against landslide danger, compacted filling can be made in front of the slope or bench slope can be made in the slope. At the same time, the natural ground is dug up and compacted according to the recompression parameters. In this case, parameters of the compacted soil can be used in stability analysis.

In order to determine the slip resistance parameters of samples taken from different points of four separate slopes, a cutting box experiment was carried out. After the experiments, c' and ϕ' values were found. The stability analyses were carried out by the model constructed and practiced in the field as shown in **Figures 8** and **9**.

The manual weight chart method was so efficient and useful in slide pattern analysis in the area as shown in **Figure 9**. The sliding surface is circular half cylindrical. The sliding mass is divided into slices as equally as possible [33].

The safety coefficient values of the landslides were calculated according to Fellenius, Bishop, and Jambu [1, 34].

Fellenius method: if the forces between slices are considered to be in the same direction but opposite and equal to each other, they are not taken into account in the analysis. In the back, only the slice weight, ground reaction, cohesion, friction resistance, and leakage forces, if any, are in balance, without drainage on partially water-saturated floors.

The rupture envelope, which is determined by the strength's test under conditions, does not parallel the normal tension axis after a point, and the ground behaves both cohesively and internally. Total tension analysis method can be applied to cover this condition and to analyze stability by sliding mass divided into a certain number of vertical slices [12, 35–37]. **Bishop method**: in this method, as in all

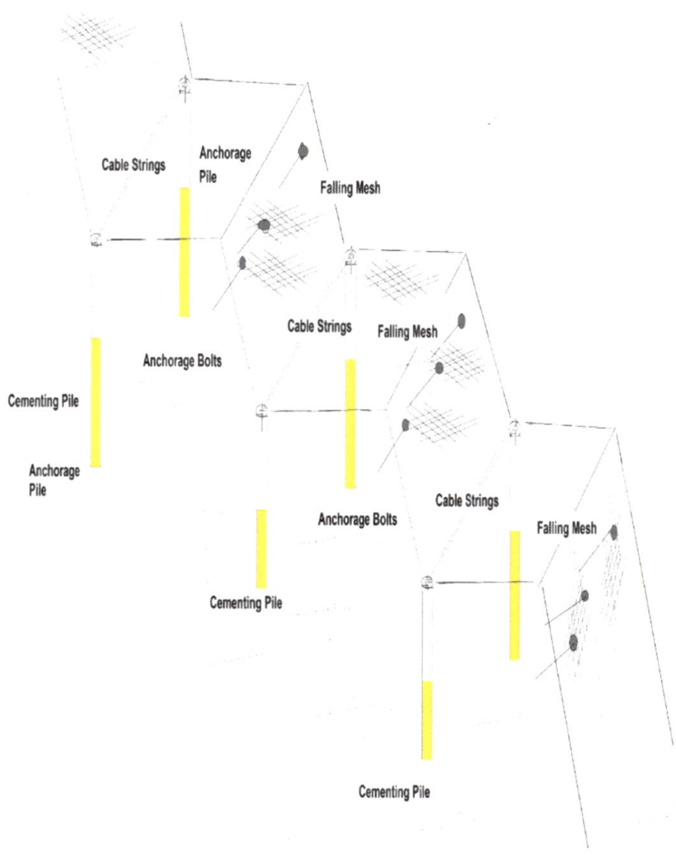

Figure 8.
North and south steep slope faces of Avgamasya No.1 pit of Şırnak asphaltite coal mine site and survey area with anchorage pile pattern and PE melting paste filling to cracks.

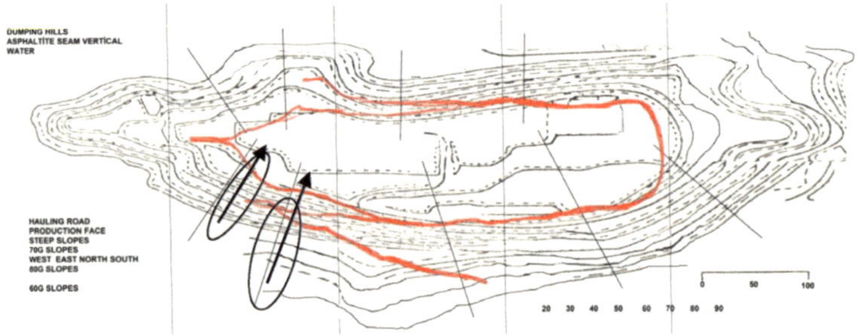

Figure 9.
Sliding shale areas contour topography of Şırnak Avgamasya open pit No.1 mining area and survey area 1/5000.

stability problems, the initial shift is taken as post-equilibrium equations are taken out as if the slope is in the limit equilibrium. Bishop performed analysis with effective stresses instead of total stress. This method is more advanced than the methods brought by Taylor and Fellenius [28, 29]. **Janbu method**: this method can be used for all types of sliding surfaces, whether circular or not. In the slope stability analysis, it is a method that takes into account the inter-slice forces for the stability analysis of the more general types of noncircular shifts with the circular shifts occurring in the homogeneous split and fillings [38].

If the properties of the ground, very weak rock mass or rust material in any slope vary frequently throughout the slope, also the applicability of circular shear analysis methods disappears due to a structural feature such as ground-rock touch or in the presence of low shear strength planar levels in the mass [12, 13, 36, 37]. Shifts under such conditions: it develops along surfaces that start circularly in noncircular or near slope-top areas and continue in planar depths [14–17]. It is a method used to examine the stability of slopes, where instabilities develop or may develop along such sliding surfaces.

4.1 Model slopes and landslide analysis

In order to evaluate geological and landslide data, 1/1000 scale topographic maps of the slopes were prepared by field studies. Polar coordinate system was used in map production and Topcon GTS 702 electronic device was used. The heights are given according to the triangulation point on the 1200 m high hill. Topographic maps prepared for 4 slopes in the study area are given in **Figure 1**. In these 4 slopes, active and potential areas in terms of stability have been distinguished as a result of the surface studies on landslide.

According to the studies conducted, the regions where landslide is developed and the areas where relative movements are observed have been determined as active landslide area. Relative movements are determined by using the stress cracks on the surface. Potential landslide areas, on the other hand, correspond to areas where there are stress cracks around these active sites, but relative movements are not currently observed. In the study area, geological sections were prepared by marking the geological outcrops on topographic sections taken from 4 different slopes.

The safety coefficients of the slopes have been used with Bishop, Janbu, and Fellenius methods and circular slip diagrams GEO5 program for different slip surfaces [39, 40]. It is taken as 1.3 in ASTM standard in the limitations of safety coefficients. In laboratory experiments, c′ and φ′ were found according to the effective and maximum resistance parameters. It is known that movements occur in the stretch cracks in the field over time. In this case, the internal parameters on the sliding surfaces are more than those in the laboratories. It will be small, in other words, it will be closer to permanent values. In this respect, the value of 1.35 was taken as the limit security coefficient.

4.2 Slope S1 landslide risk analysis in Şırnak Avgamasya open pit No.1

According to the combined ground classification (USCS) located on the skirts of Şirnak Avgamasya open pit No.1 ground, movements in the area S1 consisting of shale group stones continued as high erosion creep with steep slope. Tectonic events occurring in the region have also triggered movements and continued to this day. Nowadays, there are small-sized movements on the slope after the rains.

The discontinuous cracks were studied by cable extensometers weighted and observed as cm changes by daily periods in the control work. The slopes where landslide risk S1 is covered with pile anchorage and mesh slope stabilization by melted waste PE pressed (**Figures 8** and **10**). The maximum elevation difference between the top and heel point of the landslide S1 is 72 m, the maximum height of the slope is 80 m, and the slope angle of the slope is 58°.

In the observations made on this slope during the field studies, it was observed that it took the material that occurred in the stream and that small breaks and flows occurred after the precipitation. The geological map and landslide cross section and the surfaces where the calculations are made are given in **Figure 10**.

4.3 Rock tests and crack stabilization by melted waste PE

Mineralogical compositions of the samples were determined by means of standard chemical Ca, Mg, and silica analyzes. The samples were first brought to dimensions between 40 mm and 10 mm in jaw crusher and were homogenized by milling to 0.1 mm. Powder samples are thawed and burned with HF in platinum crucible for silica content. Chemical composition of the rocks provided in the vicinity of Şirnak province in the experiments is given in **Table 3**. The amount of silica in the marly and marly limestone was reduced.

Prior to the preparation of the melting mixture by fly ash, micropictures showed the macropores in which melted PE penetrates into the pores and makes the pores stick easily to stabilize cracks as shown under the microscope pictures (**Figure 11**).

Şirnak limestone, Midyat limestone, Şirnak marl, and marly shale were sized at 50–70 mm cubic blocks. The blocks were compressed with ELE press under 30 kN compression The blocks were crushed in test. The test results were given in

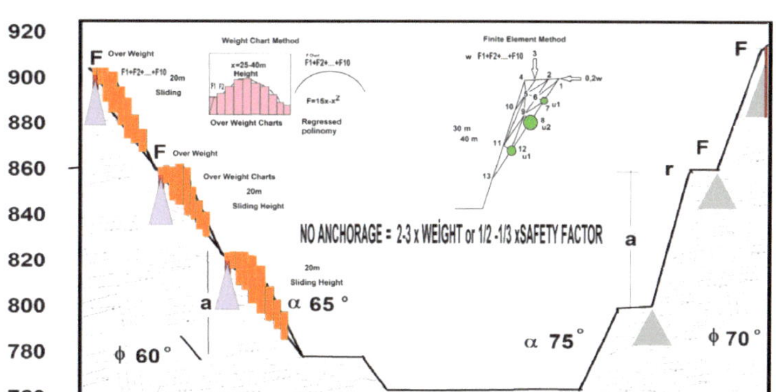

Figure 10.
Sliding shale landslide risk cross section with slope topography of Şırnak Avgamasya open pit No.1 mining, models for stability analysis.

%Component	Sırnak marly shale stone	Şırnak Marl	Şırnak porous limestone	Şırnak shale
SiO_2	9.42	24.14	2.12	48.53
Al_2O_3	6.53	12.61	1.71	24.61
Fe_2O_3	4.48	7.34	0.58	7.59
CaO	39.23	29.18	45.22	9.48
MgO	2.28	4.68	7.41	3.28
K_2O	0.53	3.32	0.40	2.51
Na_2O	0.24	1.11	0.21	0.35
Ignition loss	26.11	21.43	48.04	3.09
SO_3	0.21	0.20	0.02	0.32

Table 3.
The chemical analysis of rock specimens of Şırnak asphaltite Avgamasya open pit mine No.1 site limestone, marly shale, and shale.

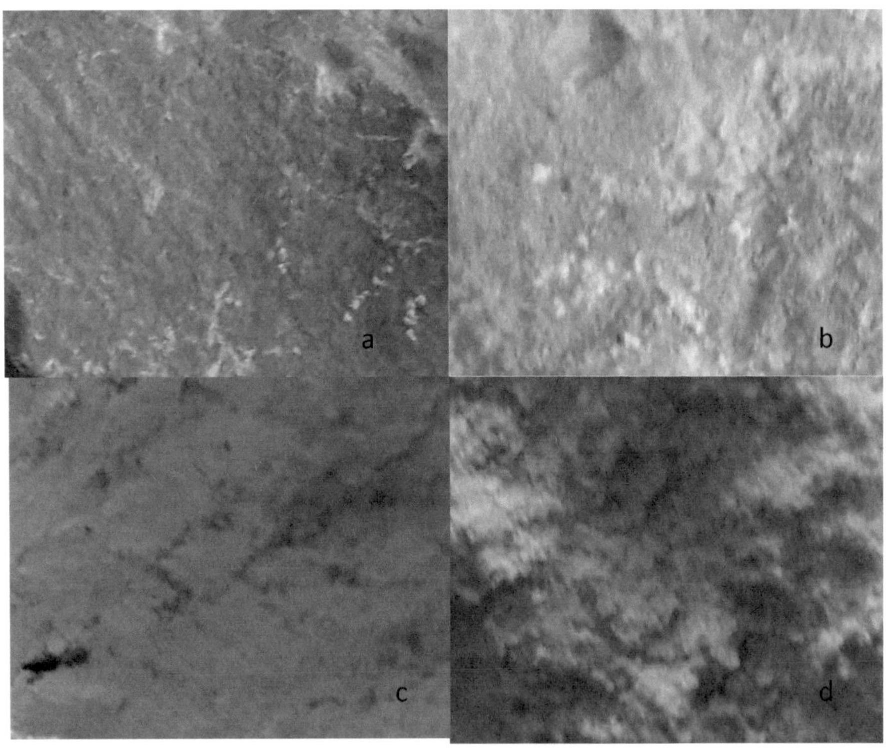

Figure 11.
(a) Shale, (b) marly shale, (c) Şırnak porous limestone, and (d) the Şırnak-altered porous limestone.

Figures 12 and **13** within the limit indentation pattern values charted, as given in **Table 3** (**Figures 14–16**).

4.4 Point loading and compressive strength tests of rocks

The test samples were produced as 5 × 5 × 5 cm blocks and 10 samples were determined to be with 95% accuracy rate by prestigious ELE brand test equipment. The results are shown in **Figures 6** and 7.

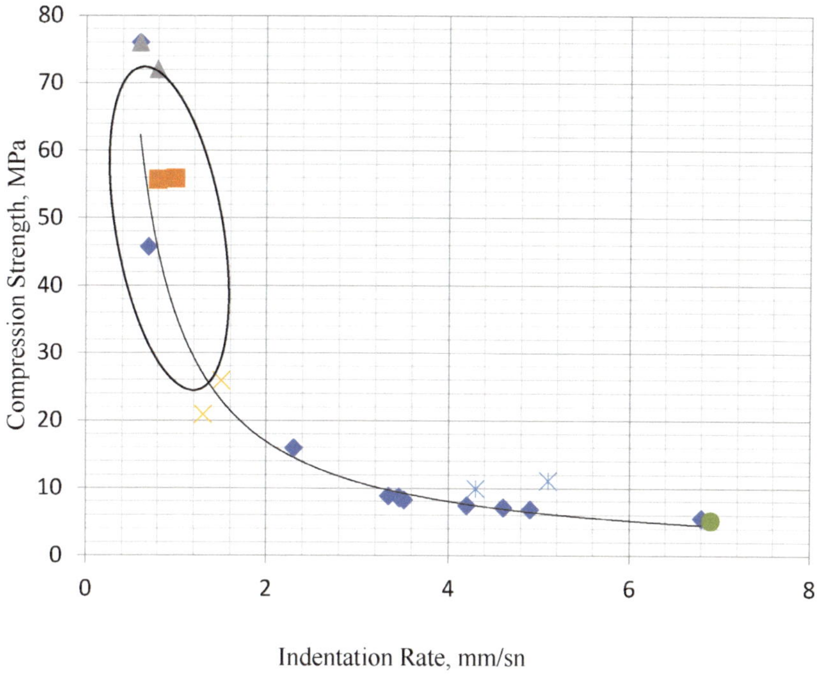

Figure 12.
Compression strength change of the Şırnak aggregates.

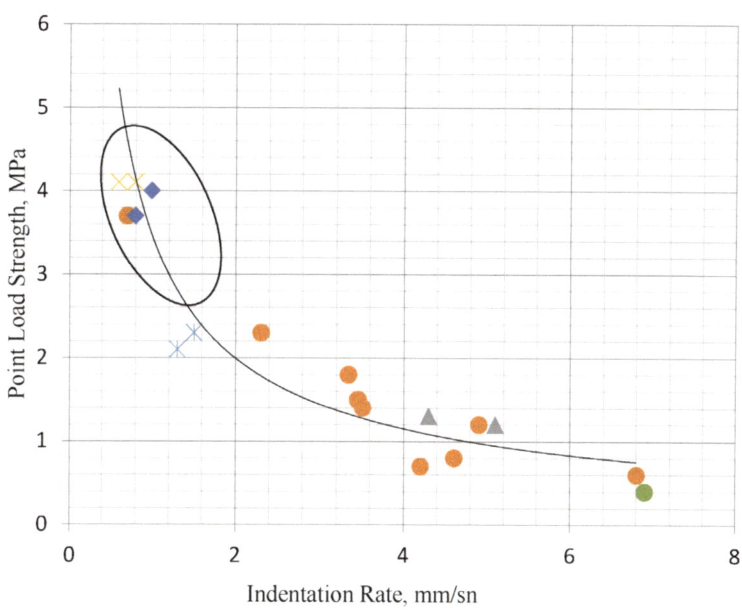

Figure 13.
Point load strength change of the Şırnak aggregates.

The rock samples have also increased the 13.6% of the pore contained in bulk pile as 25 mm fraction showed a pore rate as lower at 9.2%. The results were shown in **Figure 1**. **Figure 1** also proved that the melted interaction of 20 min is sufficient. Hence, the dissolution process reached the PE and ash saturation of the emulsified PE solution by waste machine oil. The pores in the limestone sample reached 13.6% (**Table 4**).

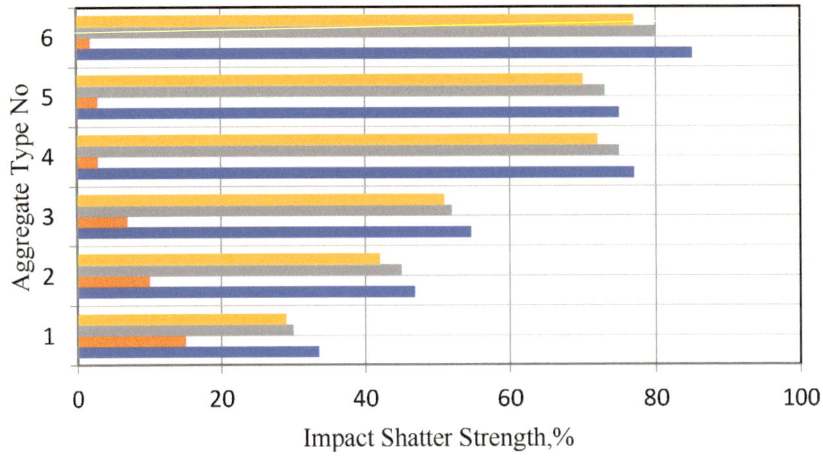

Figure 14.
Histogram view of impact, abrasion resistance: 1. Sırnak porous limestone; 2. Sırnak marl; 3. Şırnak marly shale; 4. Midyat limestone; 5. Şırnak porous limestone; and 6. Şırnak marl.

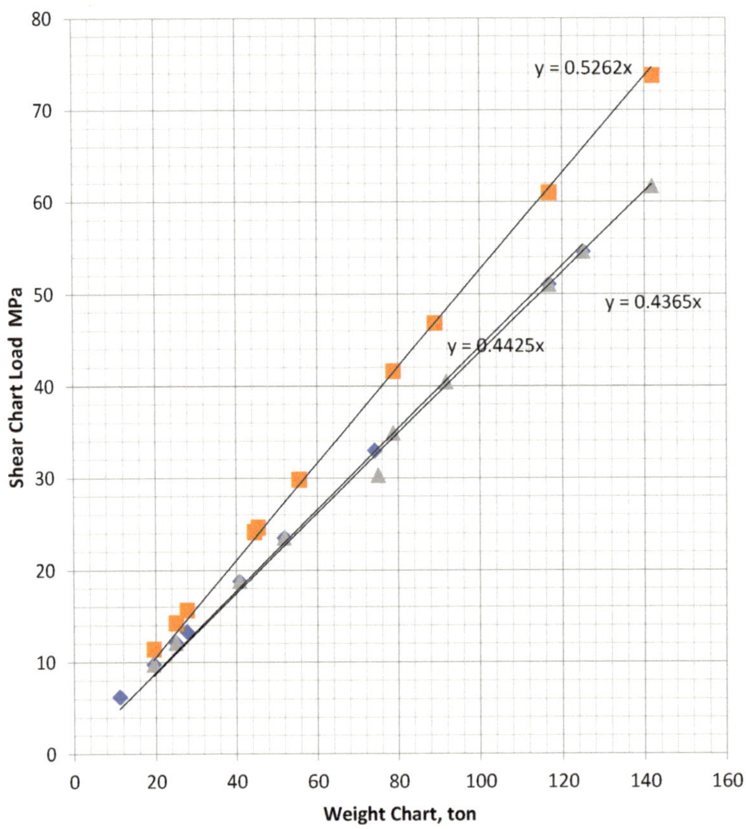

Figure 15.
Chart pattern for Workers in open pit mine excavations for S1 slope. Excavation was completed till 43° slope obtained for overburden.

Porous limestone texture, chemical interaction result, and petrographic changes are seen in **Figure 2**. It was determined that the amount of silicate of 2.3 and 4% of Şirnak altered limestone was changed and the pore size was as micro mesh with a size of 1–3 mm macro 50–300 micron. This pore amount is in microcrystalline

Figure 16.
Open pit mine No.2: excavations for steep slopes—asphaltite excavation.

%Component	Shale powder	Asphaltite slime	Tatvan pumice	Şırnak fly ash
SiO2	43.48	50.50	60.13	41.48
Al_2O3	16.10	14.61	17.22	18.10
Fe_2O3	10.52	24.30	4.59	4.52
CaO	8.48	2.30	2.48	18.48
MgO	3.80	1.28	2.17	4.20
K_2O	2.51	2.51	3.51	2.71
Na_2O	1.35	1.35	4.35	1.95
Ign.loss.	10.9	0.21	4.12	1.9
SO3	0.32	0.12	0.52	0.22

Table 4.
Chemical composition of binder fillers for crack stabilization in the Şırnak asphaltite Avgamasya open pit mine No.1.

microporous structure of 13.4–14.8% silica in Sırnak marly limestone and as 5–30 micron microporous. The degree of chemical interaction in marinated limestone and marl was not sufficient due to the silica content and the microcrystalline pore structure.

4.5 Stability risk survey for S1 slope

According to the information obtained from the drillings carried out by the Special Provincial Administration, the depth of the slope varies 11–20 m. In the investigations, the groundwater level on the slope where the landslide no S1 is located was observed at a base of 25 m. In the stability calculations made on the slope where landslide S1 developed, $c' = 190$ kg/cm^2, $\varphi' = 22°$, $\gamma_{natural} = 2.97$ g/cm^3, and $\gamma_{dry} = 2.7$ g/cm^3. Safety coefficient values of possible sliding surfaces calculated separately according to "saturated" rock were shown in **Table 6**. According to Hoek-Bray [32], the safety coefficient value is determined as 1.25. When **Table 7** is examined, since the boundary safety coefficient is taken as 1.25, it can be easily seen that the sliding slopes S1 and S2 were unstable. It is seen that the sliding surfaces 3 and 4 are very close to the limit value when it is made according to "saturated" rock and it is unstable when it is examined according to "natural."

According to the slope-clay floor combined ground classification (USCS) located in the upper skirts of the south district, the SW-SC group consists of plastic floors. Tectonic events occurring in the region have also triggered movements and continued to this day. Today, it has been observed in the field studies that there are small size movements on the slope after the rains. The slope where landslide risk S1 was developed was covered with color black showed the instability (**Figure 11**). The maximum elevation difference between the top and heel point of the landslide S1 is 75 m, the maximum height of the slope is 110 m, and the slope angle of the slope is 48°.

According to the information obtained from the drillings carried out by the Mining Bureau of Şirnak Administration Authority, the thickness of the slope varies between 11 and 20 m. In the investigations, underground water level was observed on the slope where the landslide number. In the stability calculations on the slope where the landslide S1 is developed, $c' = 199$ kg/ cm^2, $\varphi' = 22°$, $\gamma_{sat} = 2.97$ g/ cm^3, and $\gamma_{dry} = 2.77$ g/cm^3 were used. Safety coefficient values of possible sliding surfaces calculated separately according to "saturated" rocks were shown in **Table 6**. According to Hoek-Bray, the safety coefficient value is determined as 1.25. When **Tables 6** and 7 were examined and analyzed, since the boundary safety coefficient was taken as 1.25, it could be easily seen that the sliding surfaces 1 and 2 are unstable. It was seen that the sliding surfaces S3 and S4 are very close to the limit value according to saturated rock weight.

Shale rocks without drainage develop geotechnical parameters anisotropically. The slopes under compressive load show different shear strength depending on the shear face direction depending on the water content.

$$c_\theta = c_2 + (c_1 - c_2)\ \cos^2\theta \tag{1}$$

where θ is the angle made with the principal stress direction where shear stress occurs as below:

- compressive strength in the direction of θ sliding face;
- normal load stress in the vertical direction; and
- shear stress in the horizontal direction.

Compressive load values in the calculation of design model slope stability cards were given as equation below:

$$F = \sum_{0}^{i} \mathrm{N}_i \mathrm{F}_\mathrm{i} = \mathrm{N}_i \frac{C_\mathrm{i}}{\gamma H} \cos \beta^i \tag{2}$$

The compression load on the $\mathrm{N}_i$block was calculated as the ground anisotropic shale shear $\frac{C_1}{C_i}$ strength values depending on the slope angle, β. Although safety factor is designed as 1.25 and 1.35, safety factor is preferred as 1.35, according to water content greatly considered.

Due to the differences in fracture distribution, in order to determine the safe slope stability angle in the stress design cards, the sliding risk factor R_c depends on the fracture distribution percentage and angle of the slope angle in the formation, which has a high probability of slide.

The fracture or discontinuity angle t frequency% in the 20 m sliding on slope direction and the variable position in the design card $\frac{dy}{dx}$ were calculated as given in below equations and in **Tables 5–7**

$$R_c = \sum_{0}^{i} R_i F_i \tan\theta = \int_a^b e^{-ti\theta} dy^i \tag{3}$$

$$\frac{dy}{dx} = e^{-ti\theta} dy \tag{4}$$

Curing time (min)	Uniaxial compressive strength (kg/cm^2)				
Compacted binder	**1. Sample**	**2. Sample**	**3. Sample**	**4. Sample**	**Average sample**
5% fly ash					
10	205,6	205,6	209,60	203,5	204,55
30	284,5	284,5	288,4	296,5	289,80
15% fly ash					182,99
10	205,6	188,81	198,39	196,1	194,43
30	284,5	304,84	319,7	301,75	308,76
20% fly ash					203,61
10	205,6	251,17	253,9	257,56	254,21
30	284,5	360,79	369,2	360,73	363,57
25% fly ash					205,50
10	205,6	300,24	220,92	241,46	254,21
30	284,5	355,13	356,99	359,21	357,11
30% fly ash					281,89
10	205,6	323,66	316,99	313,72	318,12
30	284,5	381,17	377,42	371,8	376,80

Table 5.
Şırnak limestone aggregate sieve analysis results aggregate type.

Site	F1	F2	F3	F4	F5	F6	F7	F8	F9	F10	SF
S1	1210	2150	3270	4180	5460	6210	5740	4130	3550	2230	1.1
S2	1320	2280	3330	4270	5640	6540	5730	4580	3840	2470	1.2
S3	1110	2010	3110	4080	5200	6030	5530	3970	3110	1940	1.3
N1	1200	2100	3200	4100	5400	6200	5700	4100	3500	2200	1.2
N2	1200	2100	3200	4100	5400	6200	5700	4100	3500	2200	1.25
N3	1200	2100	3200	4100	5400	6200	5700	4100	3500	2200	1.3

5 m unit saturated weight charts kN.

Table 6.
Sliding shale landslide risk cross section with slope topography of Şırnak Avgamasya open pit No.1 mining.

Site	Joint 1,	Joint 2	Joint 3	Joint 4	Joint 5	Joint 6	Joint 7	Joint 8	Joint 9	Joint 10	Joint 11	Joint 12	Joint 13	SF 13-12
S1	321/332	515/443	527/453	418/366	546/377	621/556	574/221	513/462	655/555	223/112	418/366	513/462	655/555	1.1
S2	321/332	515/443	527/453	418/366	546/377	621/556	574/221	513/462	655/555	223/112	418/366	513/462	655/555	1.2
S3	321/332	515/443	527/453	418/366	546/377	621/556	574/221	513/462	655/555	223/112	418/366	513/462	655/555	1.3
N1	321/332	515/443	527/453	418/366	546/377	621/556	574/221	513/462	655/555	223/112	418/366	513/462	655/555	1.2
N2	321/332	515/443	527/453	418/366	546/377	621/556	574/221	513/462	655/555	223/112	418/366	513/462	655/555	1.25
N3	321/332	515/443	527/453	418/366	546/377	621/556	574/221	513/462	655/555	223/112	418/366	513/462	655/555	1.3

5 m unit saturated rock block weight points kN (H/V).

Table 7.
Sliding shale landslide risk finite analysis with slope topography of Şırnak Avgamasya open pit No.1 mining.

Safety risk parameter was calculated as 1.42 stable for 400 slopes, but for 500 and 600 slopes, the safety factors decreased to 1198 and 1060. As given in the figure, the equation slope 44.20 has been given a safety factor for a stable slope as 1320 is shown in **Figure 8**.

$$S = \frac{\Sigma(c' + \sigma' \cdot \tan \phi') \cdot \ell}{\Sigma(W \cdot \sin \alpha)} \tag{5}$$

In the observations made on this slope during the field studies, it was observed that it took the material that occurred in the stream and that small breaks and flows occurred after the precipitation. The geological map and landslide cross section and the surfaces where the calculations are made are given in **Figure 17**.

Groundwater movement change should be considered as a hazard and landslide prevention methods appropriate for the site should be determined. In addition, as the study area will be opened to urban use within the scope of the project, research and development of nonslip methods in the region is of special importance (**Tables 8–10**).

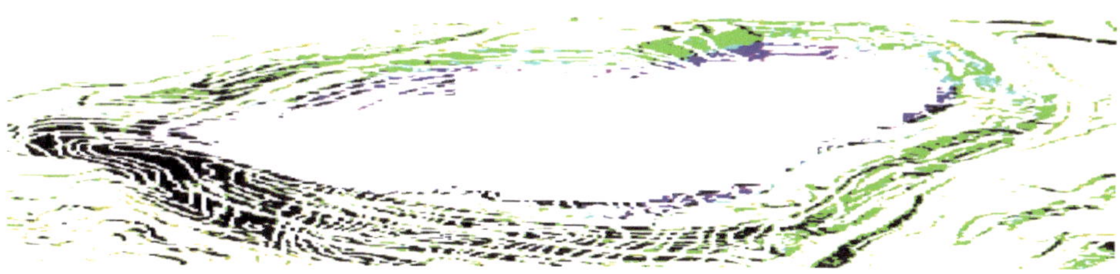

Figure 17.
Sliding shale landslide risk isocontour map; black area is unsafe area.

Chart	Block height	Block width	Block weight (ton)	Block weight (kN)	Block chart share (MPa)	Safety
1	3	4	11,12	109,0872	6,217,489	
2	5	6	27,8	272,718	13,29,372	
3	10	8	74,13,333	727,248	32,94,993	
4	15	9	125,1	1227,231	54,57,176	
5	18	7	116,76	1145,416	51,03364	
6	16	5	74,13,333	727,248	32,94,993	
7	14	4	51,89,333	509,0736	23,51,495	
8	11	4	40,77,333	399,9864	18,79,746	
9	9	3	25,02	245,4462	12,11,435	
10	7	3	19,46	190,9026	9,755,607	500
Total				5554,357	255,1988	1206

Table 8.
Weight chart calculations for S11.

Chart	Block height	Block width	Block weight (ton)	Block weight (kN)	Block chart share (MPa)	Safety
1	5	6	27,8	272,718	15,65,644	
2	7	7	45,40,667	445,4394	24,62,219	
3	12	8	88,96	872,6976	46,80,062	
4	17	9	141,78	1390,862	73,69,786	
5	18	7	116,76	1145,416	60,95,706	
6	17	5	78,76,667	772,701	41,60,992	
7	15	4	55,6	545,436	29,81,289	
8	12	4	44,48	436,3488	24,15,031	
9	9	3	25,02	245,4462	14,2408	
10	7	3	19,46	190,9026	11,40,951	500
Total				6317,967	342,9576	1418

Table 9.
Weight chart calculations for S12.

Chart	Block height	Block width	Block weight (ton)	Block weight (kN)	Block chart share (MPa)	Safety
1	9	9	75,06	736,3386	30,29,683	
2	11	9	91,74	899,9694	40,41,929	
3	15	9	125,1	1227,231	54,57,176	
4	17	9	141,78	1390,862	61,64,799	
5	18	7	116,76	1145,416	51,03364	
6	17	5	78,76,667	772,701	34,91,555	
7	14	4	51,89,333	509,0736	23,51,495	
8	11	4	40,77,333	399,9864	18,79,746	
9	9	3	25,02	245,4462	12,11,435	
10	7	3	19,46	190,9026	9,755,607	500
Total				7517,926	337,0674	1171

Table 10.
Weight chart calculations for S13.

5. Conclusions

As a result of the laboratory experiments carried out on the soil samples, it was determined that the slope material was permeable, the cohesion value ranged between 0.1 and 0.38, the internal friction angle ranged between 17.5 and 22.4, and according to the combined ground classification, the slope material consisted of shale group plastic steep slopes in the top area of low RQD. From the stability analysis performed in the light of this information, it was concluded that the slopes in S1, S2, and N1 are unstable, and the slope S3 and N2 in open pit were stable.

As a result of the field studies, it was determined that the increase in the flow rate at the peak with the melting of snow, especially in May and June, caused severe erosion on the heel of the slopes and thus had a negative effect on the stability of the slope. Chart patterns were obtained on the slopes S1, S2, and S3 to reduce sliding risk.

In order to prevent this situation with clay, it is necessary to accumulate rock material in sizes that the river cannot carry, or the creek must be improved. South and North districts of the Avgamasya open pit No.1 asphaltite mine should be reinforced in the hazardous areas according to ground water. As a result of these shock wave movements, taking precautionary measures according to the slopes and underground water floods, drainage channels should be examined again. For this reason, there is a need to investigate the stability of the slopes that will be opened to urban use.

The importance of extensometer station was very critical for over 20 mm displacement and covering the cracks by mixture of melted plastics and asphalt and facilitates the infiltration of rainwater into the channel reduces the landslide or rock flow. This drainage patterns prevents the masses creep.

The lack of vegetation in the study area prevents benefiting from these effects, which are positive in terms of stability, and hence, there is a reduction in the holding forces that keep the slopes in balance. For this reason, enrichment in terms of vegetation is an important landslide preventing parameter in the region. However, the effect of vegetation on stability will be minimal for sliding surfaces up to 30 m in depth. The weathering causes the rocks to change to a great extent, the bond between the grains to weaken and disappear completely. The rocks, which weaken as a result of weathering in the study area, are easily eroded and change the slope angles and slope altitudes. The weathering observed in the rocks in the study area also contributes negatively to the stability problems. As a result of geotechnical analysis carried out in the study area, it is concluded that very large landslides will not be expected in the future. However, this result does not eliminate the possibility of landslide danger for urban areas and urban development areas. For this reason, especially in large-scale plans such as master zoning plan and application zoning plan, geological hazard in the region should be evaluated and landslide prevention methods appropriate for the site should be determined. In addition, research and development methods to prevent instability in the region are of special importance since the study area will be opened to urban use within the scope of the project.

As a result of laboratory experiments carried out on the soil samples, it was determined that the slope material was permeable, the cohesion value ranged between 1300 and 3800 kPa, and the internal friction angle ranged between 37.5 and 22.1, and according to the combined ground classification, the slope material consisted of Shale, low marly shale, and cracked limestone low RQD group rocks. From the stability analysis performed in the light of this information, it was concluded that the slopes S1, S2, and N1 were unstable, and the slopes S3 and N2 N3 were stable.

As a result of the field investigations, it was determined that the increase in the flow rate at the peak with the melting of snow in May and June caused severe erosion on the heel of the slopes and thus had a negative effect on the stability of the slope. In order to prevent this situation, which is effective on slopes N2 and N3, it is necessary to accumulate rock material in sizes that the stream cannot bear, or the stream should be improved. Sirnak City and the surrounding area, according to Turkey Earthquake Zone Map, are located in the first degree in the danger zone. Since these regions are within the influenced area of the South East Anatolian Fault, frequent earthquakes occur in the region and some tectonic movements occur due to these earthquakes. As a result of these movements, the stability of the slopes is compromised. For this reason, there is a need to investigate the stability of the slopes that will be opened to urban use.

Plants facilitate the infiltration of rainwater into the mass and slow down and reduce superficial flow. This prevents the masses from erosion. The roots of the plants, whose roots reach deep, are mechanical.

Symbols

c' kg/cm^2	effective cohesion
c kg/cm^2	cohesion
Φ'o	effective internal friction angle
Φo	internal friction angle
τ kg/cm^2	shear stress
σ kg/cm^2	normal stress
Is	point load index
Bs	bending strength
Ps	compression strength
Wopt	optimum water content
$\gamma_{Natural}$ g/cm^3	natural unit volume weight
$\gamma_{Saturated}$ g/cm^3	saturated unit volume weight
γ_{Dry} g/cm^3	dry unit volume weight
γkmax g/cm^3	maximum dry unit volume weight
γs g/cm^3	grain unit volume weight
k	permeability coefficient
S1, S2, S3, S4, C1, C2	south and north landslide risk slopes No. 1, 2, 3, 4
S11, C11	sample taken from south and north landslide risk slopes number

Author details

Yildırım İsmail Tosun
Engineering Faculty, Mining Engineering Department, Şırnak University, Şırnak, Turkey

*Address all correspondence to: yildirimismailtosun@gmail.com

References

[1] Höek E, Bray J. Rock Slope Engineering. 3rd ed. Institution of Mining and Metallurgy: London, UK; 1981

[2] Goodman RE. Methods of Geological Engineering in Discontinuous Rocks. St. Paul, MN: West Publishing Co.; 1976. 472 p

[3] Goodman RE. Introduction to Rock Mechanics. New York: John Wiley and Sons; 1980

[4] Pells PJN. The behaviour of fully bonded rock bolts. In: Advances in Rock Mechanics. Vol. 2: Part B. Washington, DC: National Academy of Sciences; 1974. pp. 1212-1217

[5] Sjöberg J. Failure mechanism for high slopes in hard rock. In: Slope Stability in Surface Mining. Littleton, CO: Society of Mining, Metallurgy and Exploration; 2000. pp. 71-80

[6] Sjöberg J, Sharp JC, Malorey DJ. Slope stability at Aznalcóllar. In: Hustralid WA, MJ MC, DJA VZ, editors. Slope Stability in Surface Mining. Littleton, CO: Society for Mining, Metallurgy and Exploration; 2001. pp. 183-202

[7] Chen B, Liu J. Experimental application of mineral admixtures in lightweight concrete with high strength and workability. Construction and Building Materials. 2008;**22**:s.655-s.659

[8] Demirboğa R, Orung I, Gül R. Effects of expanded perlite aggregate and mineral admixtures on the compressive strength of low-density concretes. Cement and Concrete Research. 2001; **31**:1627-1632

[9] Demirdag S, Gündüz L. Strength properties of volcanic slag aggregate lightweight concrete for high performance masonry units. Construction and Building Materials. 2008;**22**:135-142

[10] Dramis F, Sorriso-Valvo M. Deep-Seated gravitational slope deformations, related landslides and tectonics. Engineering Geology. 1994;**38**:231-243

[11] Ulusoy R. Uygulamalı Jeoteknik Bilgiler. Ankara: JMO Yayınları; 1989

[12] Gündüz L. The effects of pumice aggregate/cementations on the low-strength concrete properties. Construction and Building Materials. 2008;**22**(5):721-728

[13] Gündüz L, Bekar M, Şapcı N. Influence of a new type of additive on the performance of polymer- lightweight mortar composites. Cement and Concrete Composites. 2007;**29**:594-602

[14] Gündüz L, Uğur İ. The effects of different fine and coarse pumice aggregate/ cementations on the structural concrete properties without using any admixtures. Cement and Concrete Research. 2005;**35**(9):1859-1864

[15] Gündüz L, Sarıısık A, Tozaçan B, Davraz M, Uğur İ, Çankıran O. Pumice Technology. Vol. 1. Isparta: Süleyman Demirel University; 1998. pp. 275-285

[16] Tosun Yİ, Cevizci H, Ceylan H. Landfill design for reclamation of Şırnak coal mine dumps—Shalefill stability and risk assessment. In: ICMEMT 2014, 11-12 July 2014, Prague, Czechoslovakia. 2014

[17] Tosun Yİ. A case study on use of foam concrete landfill on landslide hazardous area in Şırnak city province. In: XX Congress of the Carpathian Balkan Geological Association, Tirana, Albania, 24-26 September 2014. 2014

[18] Tosun Yİ. Shale stone and fly ash landfill use in land-slide hazardous area

in Sirnak city with foam concrete. GM Geomaterials Journal. 2014;**4**(4):141-150. DOI: 10.4236/gm.2014.44014

[19] Tosun YI. Kalker, Marn ve Şeylin Sünme Karakterizasyonu—Bitümlü Gözenekli Agrega için Don—Mikrodalga Kurutma-Bilya Darbe Dayanım Testi ile Sünme Etüdü, AGGRE 2016. In: 8th International Aggregates Symposium, October 5-7, 2016, Istanbul, Turkey. 2016

[20] Rocscience. SWEDGE—Probabilistic Analysis of the Geometry and Stability of Surface Wedges. Toronto, Canada: Rocscience Ltd.; 2001 Available from: www.rocscience.com/

[21] Rocscience Ltd. ROCLAB Software for Calculating Hoek–Brown Rock Mass Strength. Toronto, Ontario: Rocscience Ltd; 2002. Available from: www.rocscience.com/

[22] Rocscience Ltd. SLIDE—2D Slope Stability Analysis for Rock and Soil Slopes. Toronto, Ontario: Rocscience Ltd; 2002. Available from: www.rocscience.com/

[23] Pritchard MA, Savigny KW. Numerical modelling of toppling. Canadian Geotechnical Journal. 1990;**27**: 823-834

[24] Pritchard MA, Savigny KW. The Heather Hill landslide: an example of a large scale toppling failure in a natural slope. Canadian Geotechnical Journal. 1991;**28**:410-422

[25] Sonmez H, Ulusay R. Modifications to the geological strength index (GSI) and their applicability to the stability of slopes. International Journal of Rock Mechanics and Mining Sciences. 1999; **36**(6):743-760

[26] Sageseta C, Sánchez JM, Cañizal J. A general solution for the required anchor force in rock slopes with toppling failure. International Journal of Rock Mechanics and Mining Sciences. 2001; **38**:421-435

[27] Anbalagan R. Landslide hazard evaluation and zonation mapping in mountainous terrain. Engineering Geology. 1992;**32**:269-277

[28] Anonymous, a. GEO5—Engineering Manuals—Part 1, Part 2. 2011. Available from: http://www.finesoftware.eu/geotechnical-software/

[29] Anonymous, b. GEO5—FEM—Theoretical Guide. 2011. Available from: http://www.finesoftware.eu/geotechnical-software/

[30] Anonymous, c. Türkiye Deprem Bölgeleri Haritası. Ankara: Afet ve Acil durum Yönetimi Başkanlığı Deprem Dairesi Başkanlığı; 2012

[31] Duncan JM. Landslides: Investigation and mitigation - Chapter 13. In: Soil Slope Stability Analysis. Washington, DC: Transportation Research Board; 1996. pp. 337-371. Special Report 247

[32] ASTM. Standard Test Method for Direct Shear Test of Soils under Consolidated Drained Condition1990. pp. D3080-D3090

[33] Dershowitz WS, Einstein HH. Characterizing rock joint geometry with joint system models. Rock Mechanics and Rock Engineering. 1988;**20**(1):21-51

[34] Höek E. Estimating the stability of excavated slopes in opencast mines. Institution of Mining and Metallurgy. 1970;**A105**:A132

[35] Fell R. Landslide risk assessment and acceptable risk. Canadian Geotechnical Journal. 1994;**31**:261-272

[36] Erdoğan TY. Beton. Ankara: ODTÜ Geliştirme Vakfı Yayıncılık ve İletişim A.Ş; 2003

[37] Gündüz L. Use of quartet blends containing fly ash, scoria, perlitic pumice and cement to produce cellular hollow lightweight masonry blocks form on-load bearing walls. Construction and Building Materials. 2008;**22**:747-754

[38] Ulusoy R. Şev Stabilite Analizlerinde Kullanılan Pratik Yöntemler ve Geoteknik Çalışmalar. MTA Yayınları, Eğitim Serisi; 1982. p. 25

[39] Görög P, Török Á. Slope stability assessment of weathered clay by using field data and computer modelling: a case study from Budapest. Natural Hazards and Earth System Sciences. 2007;**7**:417-422. Available from: www.nat-hazards-earth-syst-sci.net

[40] Görög P, Török Á. Stability Problems of Abandoned Clay Pits in Budapest, IAEG2006 Paper Number 295. London: The Geological Society of London; 2006

Chapter 8

Specific Solution of Deformation Vector in Land Subsidence for GIS Applications to Reclaiming the Abandoned Magnesite Mine in the East of Slovakia

Vladimír Sedlák

Abstract

Mining activity influences on the environment belong to the most negative industrial influences. Mining subsidence on the earth surface is a result of underground mining. The present study deals with the theory of specific procedures for solving the deformation vector in the case of an objective disturbance of data homogeneity in the geodetic network structure of the monitoring station in monitoring mining subsidence. The theory was developed for the mining subsidence created on the earth surface of the mining landscape, where the abandoned magnesite mine Košice-Bankov in the East of Slovakia was operated for many decades in the twentieth century. The achieved results and outputs were implemented into the GIS tools for the plan of the process of gradual reclaiming the entire mining landscape of Košice-Bankov. The aim of the deformation measurements was to determine the exact boundaries of the subsidence edges with the residual movement zones for the purpose of comprehensive reclaiming the devastated mining landscape. Some numerical and graphical results from the deformation vectors survey in the abandoned magnesite mine Košice-Bankov are presented. The obtained results in GIS were supplied for the needs of the Municipality of the city of Košice to the realization of the reclaiming work.

Keywords: subsidence, deformation vector, geodetic network, Gauss-Markov model, test statistics, reclaiming

1. Introduction

At present, with an extremely sharp increase in people's material needs, priority must be given to their security from any economic prosperity of many countries around the world. As the extraction and processing mineral resources increase, so does the economic prosperity of the country. To protect the environment, which should be an intact ecosystem, it is necessary to protect the lives of people and their property from adverse industrial impacts. One of the most negative industrial impacts on the whole ecosystem is the adverse impact of any mining activity. The land subsidence (mining subsidence, hereinafter referred to as subsidence) is created on the earth surface as a result of underground extraction of the mineral deposits especially at the chamber mining by a caving method [1, 2]. In many cases, the mine subsidence represents the large-scale and very deep downthrown blocks that is dangerous for any movement of people in them [1–7]. As at the deep mining there are many large voids created in the rock massif (especially during the aforementioned mining by the chamber method), their collapse occurs with the manifestations of deformations on the earth surface mostly in the form of subsidence. The collapse of these cavities can occur at different time horizons, i.e., from the commencement of mining up to several years, or even decades or more after the end of mining. Especially for endangering the lives of people and their property, the most dangerous are the sudden and unexpected formation of invasions of the earth surface over the mined rock massif, which often happens even in some abandoned mines [6, 8, 9].

In the sense of several scientific studies, as well as theoretical and practical knowledge on mining impacts on the earth surface, it is very difficult to accurately predict how long the movements in the subsidence will be in existence and when these movements will be ended [1, 2, 10–12]. Current statistics resulting from many deformation measurements in the subsidence confirm that around 60–90% of total subsidence movements occur within the first few weeks up to months after the excavation of the underground space in the rock massif. The remaining movements in the subsidence persist even after the end of mining, and at a decreasing rate these movements can last for 3–5 years and longer and in rare cases even several decades.

In order to protect the lives of people and their property, as well as the complex environment, it is essential to constantly investigate any movements in the subsidence on the earth surface even after mining has ended [2, 13, 14]. The most investigated movements in the subsidence with their prediction based on three-dimensional (3D) modeling are on coal fields [3, 5, 11, 12, 14–17].

The nature and magnitude of the subsidence on the earth surface depends mainly on tectonic and geological conditions and also on the overload of the rock massif above the excavated space. Knowing the range, i.e., localization of the edges (boundaries) of the subsidence in mining territories, may provide for more precise placement of technical barriers (fencings and warning boards) and thus help to prevent persons and animals from entering these danger zones. Geodetic methods for investigation of deformation vectors, which can be derived from the processing of some specific geodetic measurements at the monitoring stations based on these mining territories, are the priority methods for determining the extent of movement in the subsidence on the earth surface above the excavated underground space. The deformation vectors in their 3D model concept most markedly characterize any movements of the earth surface, buildings, and other civil engineering structures located in the mining territory with the occurrence of the subsidence. In many cases, 3D modeling of the deformation vectors is based on regular (periodic) monitoring of the spatial changes in suitably structured points of the geodetic network of a monitoring station located on the earth surface or on buildings and civil engineering structures. Such periodic monitoring of the spatial changes of 3D coordinates of the geodetic network points is mostly realized by the application of classical geodetic terrestrial methods (tacheometry, trigonometry, traverse surveying, leveling, etc.), which are used extensively up to now or are currently increasingly using advanced methods of the satellite technologies within the Global Navigation Satellite Systems (GNSS) based on the US Global Positioning System (GPS), Russian Global Navigation Sputnik System (GLONAS), European Galileo, and Chinese Compass (BeiDou II). At present, very specific measurement technologies such as Interferometric Synthetic Aperture Radar (InSAR), i.e., the radar surveying technology, and Light Detection and Ranging (LiDAR), i.e., the laser surveying technology, are increasingly being applied to deformation measurements [14, 15, 18–26]. The deformation vectors are the outputs of 3D data processing from the abovementioned geodetic measurements of deformations (deformation measurements) of the earth surface or building structure objects, but several geodetic measurements are combined and thus confirmed by some physical measurements based on the measurement of stress and strain states. By their size and position in 3D space, deformation vectors can provide a global overview not only of the nature of the current state of deformation of the monitoring subsidence in the mining area, but also of the nature of the further development of these deformations with the required time predictions [1, 2, 10, 19]. At the same time, when monitoring particularly important buildings located close to the subsidence, additional geophysical measurements of stress–strain states of the rock massif must be carried out underground (in the extracted mining space). Achievement of the unstressing states in the rock massif above the extracted space can be realized in many cases by the so-called destress blasting [27, 28].

Periodically repeating measurements on parts respectively on the whole geodetic network of some monitoring stations to measure deformations of the earth surface (e.g., subsidence) or construction objects and engineering structures can sometimes be complicated for objective reasons. Most of the monitoring stations consist of a network of firmly stabilized points (geodetic network) on the surveyed earth surface or on surveyed buildings and other engineering structures. By various geodetic methods, measurements are made on these points of the geodetic network to determine their 3D positions (i.e., coordinates X, Y, and Z) in the given Cartesian coordinate system [5, 14]. During the realization of long-term (multiyear or even several decades) geodetic measurements, undesirable and unpredictable obstructions and interventions into the monitoring station and so into the geometric structure of the geodetic network may also occur. In many cases, such undesirable interventions into the structure of the geodetic network include predominantly loss or damage to points due to uncontrolled earthworks or construction work in the territory of the monitoring station. Due to such undesirable interference with the monitoring station, the geometric and data structure of the points of the geodetic network is not homogeneous during all periodic deformation measurements.

This means that the measured elements of the applied geodetic measurements at the points of the geodetic network of the monitoring station at individual time epochs are no longer identical and cannot be maintained as they were at the time of the initial (primary, i.e., zero) measured epoch. When changing the geometrical structure of the points of the geodetic network, it is no longer possible to maintain, in particular, the spatial configuration of the measuring sights between the determined points of the network. Disturbance of stability or destruction and the loss of multiple points of the geodetic network also results in a disturbance of the geometric and thus data structure of the whole network of the monitoring station. In such cases, the re-stabilization of destroyed or lost points (in particular object points) at the original or other places of the geodetic network will not help either. Nor will it help to replace the original measured variables (which are no longer measurable in the subsequent monitoring epochs due to damage of the geodetic network points) by other measured variables [1, 10, 25, 29].

In the evaluation of deformation vectors and their 3D modelling, the time factor of the gradual creation of the subsidence over the mined-out space underground plays an important role in its overall evaluation of the earth ground movements.

The possibility in improving polynomial modeling of the subsidence is conditioned by the knowledge to detect position of the so-called breakpoints, i.e., the points in the surface in which the subsidence border with a zone of breaches and bursts start to develop above the mined mineral deposit. This means that the

breakpoints determine the edges of the subsurface at which the naturally consistent coherent of the earth surface is broken and the subsidence begins to form.
3D deformation vector models help to support the location of the breakpoints [10–13, 18, 21].

2. Theory of the deformation vector specific solution

As already mentioned in the Introduction, the geometric and thus data structure of the geodetic network of the monitoring station in the subsidence may be changed by some external intervention, such as some unforeseen earthworks and construction works at the monitoring station. Estimation of the structures of geodetic networks based on the Gauss-Markov model is the most used and the most effective method for their processing. In determining the statistical formulation of the Gauss-Markov model, we start from the following equations [11, 12, 17, 30–33]:

$$\boldsymbol{v} = \boldsymbol{A}\left(\hat{\boldsymbol{C}} - \boldsymbol{C^o}\right) - \left(\boldsymbol{L}_{(0)} - \boldsymbol{L^o}\right) = \boldsymbol{A}d\hat{\boldsymbol{C}} - d\boldsymbol{L} \tag{1}$$

$$\boldsymbol{\Sigma}_L = \sigma_0^2 \boldsymbol{Q}_L \tag{2}$$

where $\boldsymbol{v}$ is the vector of corrections of the measured (observed) values $\boldsymbol{L}$,$\boldsymbol{A}$ is the configuration (modeling) matrix of the geodetic network (otherwise called also the Jacobian matrix), i.e., the matrix of the partial derivatives of functions $\boldsymbol{L^o} = f(\boldsymbol{C^o})$ by the vector $\boldsymbol{C^o}$,$\hat{\boldsymbol{C}}$ is the vector of the aligned 3D coordinate values, $\boldsymbol{C^o}$ is the vector of the approximate 3D coordinate values, $\boldsymbol{L}_{(0)}$ is the vector of the approximate observation magnitude values of the observed elements in the first measuring epoch $t_{(0)}$, $\boldsymbol{L}^o$ is the vector of the approximate observation magnitude values of the observed elements, $d\hat{\boldsymbol{C}}$ is the deformation vector, $d\boldsymbol{L}$ is the vector of the measured values supplements, $\boldsymbol{\Sigma}_L$ is the covariance matrix of the measured values, σ_0^2 is a priori variance, and $\boldsymbol{Q}_L$ is the cofactor matrix of the observations.

It will also be appeared in the changed structures, let us say in a size of the matrixes and vectors $\boldsymbol{A}, \boldsymbol{Q}_L, \boldsymbol{C^o}$ and $\boldsymbol{L^o}$. These matrixes and vectors enter into the presupposed model of a network adjustment following out from the Gauss-Markov model [11, 12, 29–31].

2.1 Deformation vector

If between monitoring epochs there are no changes in the geometrical and observational structure of the geodetic network, then the matrixes and vectors $\boldsymbol{A}, \boldsymbol{Q}_L, \boldsymbol{C^o}$and $\boldsymbol{L^o}$ remain identical for each epoch. Only in such case the deformation vector $d\hat{\boldsymbol{C}}$ can be determined by a conventional procedure according to the following model [2, 16, 17]:

- In the basic (first) monitoring epoch $t_{(0)}$, we have the vector $\hat{\boldsymbol{C}}_{(0)}$ of the adjusted 3D coordinates of the observed points which are obtained according to the Gauss-Markov model:

$$\hat{\boldsymbol{C}}_{(0)} = \boldsymbol{C^o} + \left(\boldsymbol{A}^T\boldsymbol{Q}_L^{-1}\boldsymbol{A}\right)^{-1}\boldsymbol{A}^T\boldsymbol{Q}_L^{-1}\left(\boldsymbol{L}_{(0)} - \boldsymbol{L^o}\right) = \boldsymbol{C^o} + \boldsymbol{G}\left(\boldsymbol{L}_{(0)} - \boldsymbol{L^o}\right) \tag{3}$$

- In the other following epochs $t_{(i)}$, we also obtain the vector $\hat{\boldsymbol{C}}_{(i)}$ of the adjusted 3D coordinates of the observed points according to the equation:

$$\hat{C}_{(i)} = C^o + \left(A^T Q_L^{-1} A\right)^{-1} A^T Q_L^{-1} \left(L_{(i)} - L^o\right) = C^o + G\left(L_{(i)} - L^o\right) \tag{4}$$

- Thus, for the deformation vector $d\hat{C}$ will be valid in the following equation:

$$d\hat{C} = \hat{C}_{(i)} - \hat{C}_{(0)} = G\left(L_{(i)} - L^o\right) \tag{5}$$

where $L_{(0)}$ and $L_{(i)}$ are the vectors of the observed magnitude values in the epochs $t_{(0)}$ and $t_{(i)}$.

Furthermore, we consider the case when there is a change in the geometric and thus in the data structure of the geodetic network of the monitoring station between the individual epochs of measurements. It means that the geometric and data structure of the geodetic network between the basic epoch $t_{(0)}$ and the actual epoch $t_{(i)}$ is changed. Then the original matrixes and vectors A, Q_L, C^o and L^o are transformed into the following equations:

$$\overline{A} = A + dA \tag{6}$$

$$\overline{Q}_L = Q_L + dQ_L \tag{7}$$

$$\overline{C}^o = C^o + dC^o \tag{8}$$

$$\overline{L}^o = L^o + dL \tag{9}$$

According to Eqs. (6)–(9), the vectors $\hat{\overline{C}}_{(0)}$ and $\hat{\overline{C}}_{(i)}$ of the adjusted 3D coordinates of the observed points in the epochs $t_{(0)}$ and $t_{(i)}$ will be determined:

$$\hat{\overline{C}}_{(0)} = \overline{C}^o + \left(\overline{A}^T \overline{Q}_L^{-1} \overline{A}\right)^{-1} \overline{A}^T \overline{Q}_L^{-1} \left(L_{(0)} - \overline{L}^o\right) = \overline{C}^o + \overline{G}\left(L_{(0)} - \overline{L}^o\right) \tag{10}$$

$$\hat{\overline{C}}_{(i)} = \overline{C}^o + \left(\overline{A}^T \overline{Q}_L^{-1} \overline{A}\right)^{-1} \overline{A}^T \overline{Q}_L^{-1} \left(L_{(i)} - \overline{L}^o\right) = \overline{C}^o + \overline{G}\left(L_{(i)} - \overline{L}^0\right) \tag{11}$$

and then the deformation vector $d\hat{C}$ is expressed according to Eq. (5) in the form

$$d\hat{\overline{C}} = \hat{\overline{C}}_{(i)} - \hat{\overline{C}}_{(0)} \tag{12}$$

which not only expresses the 3D changes in the coordinates of the geodetic network points between the individual epochs of measurement, but such deformation vector can also express changes in the overall structure (geometric and data structure) of the geodetic network. The deformation vector $d\hat{\overline{C}}$ thus obtained will not provide reliable data for testing the particular deformations in the subsidence.

The proposed and presented theory of the specific solution of the deformation vector in the case of any structural changes in the geodetic network will be acceptable for its proving in an analytical way, if we compare the deformation vector structures $d\hat{C}$ and $d\hat{\overline{C}}$ expressed according to Eqs. (5) and (12). Then the structure of the deformation vector $d\hat{\overline{C}}$ is expressed according to Eq. (12), and the further equation will be valid:

$$\begin{aligned} d\hat{\overline{C}} &= \left[\overline{C}^o + \overline{G}\left(L_{(i)} - \overline{L}^o\right)\right] - \left[C^o + G\left(L_{(0)} - L^o\right)\right] \\ &= \overline{G}\left(L_{(i)} - L^o\right) - G\left(L_{(0)} - L^o\right) + \overline{C}^o - C^o \end{aligned} \tag{13}$$

and on the base of Eqs. (6)–(9) and also on the base of the linearization of $\overline{G}$ into $\overline{G} = G + dG$, the following derivation will be applied for the deformation vector $d\hat{\overline{C}}$:

$$d\hat{\overline{C}} = (G + dG)\left(L_{(i)} - \overline{L}^o\right) - G\left(L_{(0)} - L^o\right) + dC^o = \overline{G}\left[L_{(i)} - (L^o + dL^o)\right] + dG\left(L_{(i)} - \overline{L}^o\right) - \\ -G\left(L_{(0)} - L^o\right) + dC^o = G\left(L_{(i)} - L^o\right) + GdL^o + dG\left(L_{(i)} - L^o\right) - G\left(L_{(0)} - L^o\right) + dC^o = \\ = G\left(L_{(i)} - L_{(0)}\right) + GdL^o + dG\left(L_{(i)} - \overline{L}^o\right) + dC^o \tag{14}$$

and finally the deformation vector $d\hat{\overline{C}}$ will be calculated according to the following equation:

$$d\hat{\overline{C}} = d\hat{C} + \delta d\hat{C} \tag{15}$$

Eq. (15) notates that the deformation vector $d\hat{\overline{C}}$ (calculated at some changes in the geodetic network structure) is different from its vector of the correct values $d\hat{C}$ only by the component $\delta d\hat{C}$ (i.e., the correction component of the deformation vector corrections). In such a set case, the component $\delta d\hat{C}$ is generated not only by the spatial movement of points in the geodetic network between the particular epochs of the geodetic measurements, but at the same time it is generated by changes in the geometric and data structure of the network between the particular epochs due to some changes in its point field.

To avoid the so-called degradation of the deformation vector $d\hat{C}$ due to changes in the geometric and data structure of the geodetic network and at the same time for the deformation vector to express the real spatial changes in the subsidence, the presented theory offers the following procedures:

- The geodetic networks at the monitoring stations shall be designed in order to achieve the maximal physical integrity of its points (object and especially reference points) throughout the entire monitoring period. When designing a monitoring station, expert consultation with representatives of a spatial planning and also with the mine district owners is essential.

- If some reference points were lost or destroyed, new points should be stabilized in enough proximity to these lost or destroyed reference points as possible. The same principle is held for the object points.

- However, if the matrixes A and Q_L are significantly or even slightly changed between the monitoring epochs $t_{(0)}$ and $t_{(i)}$ (e.g., in $t_{(0)}$ the geodetic network was measured by a trilateration measurement way, and in $t_{(i)}$ by traverse measurement way, it is necessary to observe (measure) other new magnitudes, etc.), then the deformation vector $d\hat{C}$ can be determined according to the following equations:

$$d\hat{C} = C^o + \left(A^T Q_L^{-1} A\right)_{(i)}^{-1} A_{(i)}^T Q_{L(i)}^{-1}\left(L_{(i)} - L^o\right) \\ - \left[C^o + \left(A^T Q_L^{-1} A\right)_{(0)}^{-1} A_{(0)}^T Q_{L(0)}^{-1}\left(L_{(0)} - L^o\right)\right] \tag{16}$$

and

$$d\hat{C} = G_{(i)} L_{(i)} - G_{(0)} L_{(0)} - L^o\left(G_{(i)} - G_{(0)}\right) \tag{17}$$

because using the identical $\boldsymbol{C^o}$ and $\boldsymbol{L^o}$ are not the problem to adhere in the individual epochs. Or the deformation vector corrections $\delta d\hat{\mathbf{C}}$ are calculated according to Eqs. (10), (11), and (13), so that the deformation vector $d\hat{\mathbf{C}}$ is then corrected according to the introduced Eq. (15).

3. Study case example

3.1 Study region description

The monitoring station is situated in the territory of the mining field of the abandoned magnesite mine of Košice-Bankov. This territory was characterized by a devastated mining surface with many mining tailing piles and especially the large subsidence. The city district of Košice-Bankov is located in the northern part of the city of Košice. In addition to the abandoned magnesite mine, there is a very popular urban recreational and touristic resort, located in the large urban forest park of the city of Košice. The territory of the urban recreational and touristic resort and forest park are situated in close proximity, respectively, in the territory above the mining field of the former magnesite mine (**Figure 1**).

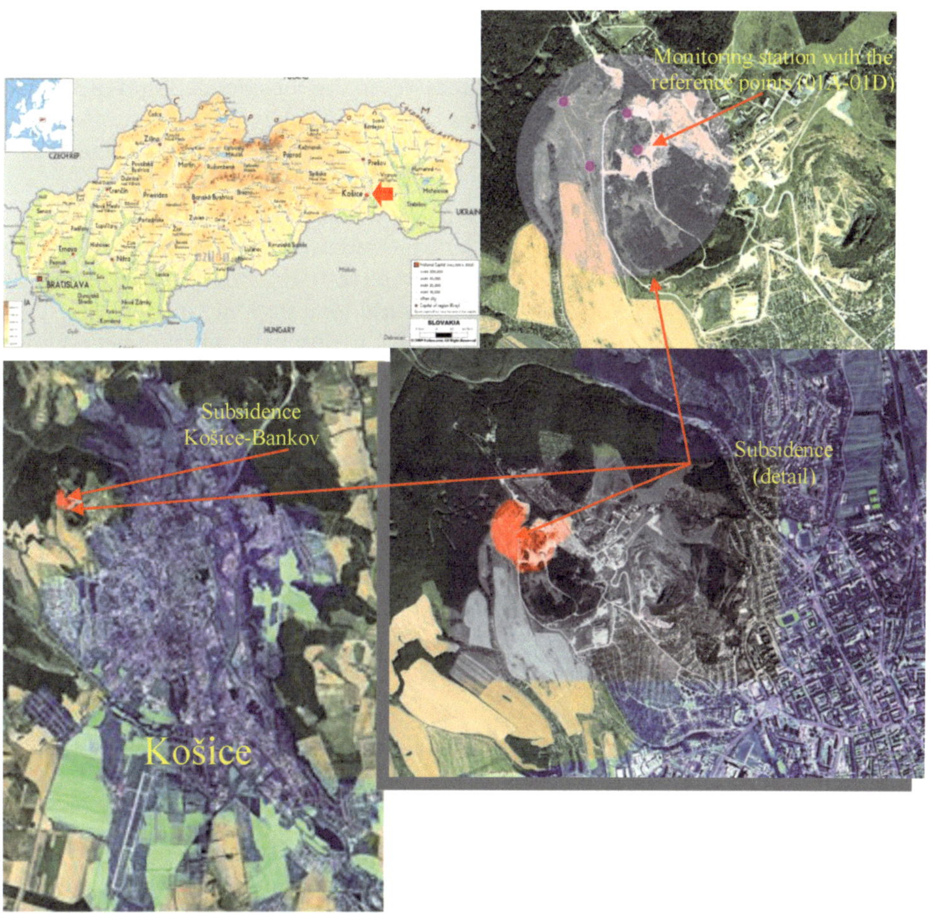

Figure 1.
Ortho-photo map of the city of Košice with a detail view to the mine field of Košice-Bankov.

Until the 1970s of the twentieth century, any systematic attention was not paid to the extent of mining damage on the earth surface at the territory of Košice-Bankov as a result of magnesite mining. It was only after this period that scientific studies began to be taken into account when dealing with the creation of the large subsidence and devastation of the protected area of the forest park and the environmental protection in the tangent territory of Košice-Bankov. The gradual development of the subsidence in the mining area of the magnesite mine of Košice-Bankov in the east of Slovakia has been monitored since the end of the 1970s by systematic geodetic measurements. The monitoring station project to monitor the development of the subsidence in the territory of Košice-Bankov was designed and implemented by researchers of the Technical University of Košice in 1976, when the first geodetic measurements were carried out and later by researchers of the Pavol Jozef Šafárik University in Košice. The first observed data were obtained from this monitoring station in autumn 1976, and since then the regular periodic spring and autumn geodetic terrestrial and satellite (GPS, GNSS) measurements were performed every year.

Before reclaiming the mining landscape on the territory of Košice-Bankov, the monitoring station was located on the site of the former subsidence at the mining shaft, which was called the Western shaft. The monitoring station was built from the geodetic network consisting of the network of the reference points (No: 01A, 01B, 01C, 01D) and the network of the object points (78 points in total). The object points were geometrically grouped into six geodetic network profiles (0–V) (**Figure 2**). All geodetic network profiles of the monitoring station of Košice-Bankov were geometrically spaced across and along the expected movements in the subsidence (**Figure 2**). Gradually, by creating the subsidence, some object points were destroyed by the nature destructive processes in the subsidence. **Figures 3** and **4** show the panoramic views to the subsidence of Košice-Bankov from the southwestern edge of this subsidence in 2001 and 2002. In that time the magnesite mine had been out of its operation for 3 years.

3D data (elements) of geodetic network points of the monitoring station were initially (since 1076) measured by 3D geodetic measurements (position and leveling measurements) by the application of the classical terrestrial geodetic technologies using the optical geodetic theodolites, electro-optical total stations, and leveling devices for a very precise leveling. Later, since 1977, the periodic measurements at the monitoring station have been made by the satellite geodetic methods GPS and GNSS, i.e., Trimble 3303DR Total Station, GPS: ProMark2, and GNSS: Leica Viva GS08.

In both geodetic periodic measurement technologies (terrestrial and satellite), regular geodetic measurements were performed twice a year, i.e., during the spring and autumn months [12, 17]. In 1981, some of the object points of the geodetic network of the monitoring station were damaged, respectively; the points were completely destroyed by some unplanned and uncontrolled earthworks in the vicinity of the subsidence (points No. 2, 3, 30, 38, 104, and 105 and 227[1] on the profiles No. 0, I, and II). Most of the damage or destruction of the abovementioned points occurred during the adjustment of some forest stands in the nearby forest park and by some earthworks on the surrounding mining tailing piles.

3.2 Accuracy and quality assessment of the geodetic network

1D, 2D, and 3D accuracy of the geodetic network points of the monitoring station was evaluated by testing the global and local network indices. The global indices were numerically expressed to assess the accuracy of the entire geodetic network.

[1] The reference point No. 227 (profile II) was rebuilt instead of the point No. 226.

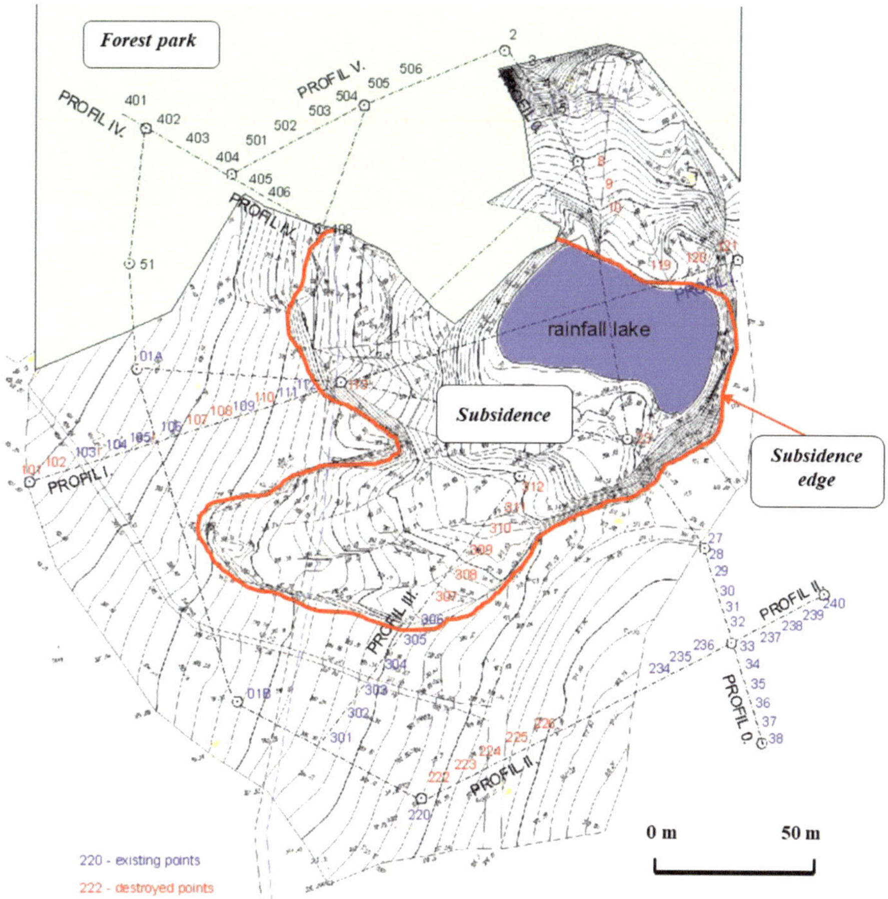

Figure 2.
The monitoring station of Košice-Bankov (reference points 01C and 01D destroyed points).

Figure 3.
The subsidence of Košice-Bankov; panoramic view: Autumn 2001.

For the global indices, the following were tested: $tr\ (\boldsymbol{\Sigma}_{\hat{C}})$, i.e., the track of the covariance matrix $\boldsymbol{\Sigma}_{\hat{C}}$, the volume global indices, and $\det(\boldsymbol{\Sigma}_{\hat{C}})$, i.e., the determinant of the covariance matrix.

In fact, the local indices were the point indices that characterize the reliability of the geodetic network points: the mean 3D error $\sigma_p = \sqrt{\sigma^2_{\hat{X}_i} + \sigma^2_{\hat{Y}_i} + \sigma^2_{\hat{Z}_i}}$, the mean

Figure 4.
The subsidence of Košice-Bankov; panoramic view: Spring 2002.

coordinate error $\sigma_{XYZ} = \sqrt{\frac{\sigma^2_{\hat{X}_i} + \sigma^2_{\hat{Y}_i} + \sigma^2_{\hat{Z}_i}}{3}}$, and the confidence absolute ellipses or ellipsoids, which were used to assess the real 2D or 3D point accuracy. It is necessary to know the design elements of the ellipse of errors, i.e., the semi-major axis a, the semi-minor axis b, the bearing φ_a of the semi-major axis, and the ellipsoid flattening f ($f = 1 - b/a$).

The geodetic network quality characteristics to be assessed are, above all, the accuracy and reliability of the position of the points. In addition to numerical expressions, the accuracy of the point position can also be expressed using the graphical indicators such as the reliability curves and the ellipse of confidence (ellipsoids of reliability in the case of 3D space). The ellipsoids determine the random space in which the actual location of the points will lie with a probability of $1\text{-}\alpha$, where α is the level of significance chosen according to which the ellipsoids are of different size. For 3D space in a geodetic practice, the standard confidence ellipsoids are usually used. The design parameters of such ellipsoids can be derived either from the cofactor matrix $\mathbf{Q}_L$ of the adjusted coordinates, where the design parameters are arranged on the main diagonal of the matrix, or from the covariance matrix of the coordinate estimations $\mathbf{\Sigma}_{\hat{C}}$ of the determined points, which are arranged also on the main diagonal of that matrix.

All calculated data according to the submitted specific theory of the deformation vector solution in the case of any geometric and data changes in the geodetic network of the monitoring station are shown in **Tables 1–5** which are focused on the accuracy and quality assessment of the network in the years 1976 and 2014 (1976/2014[2]) (**Tables 1–5**). **Tables 1–5** comprehends the adjusted mean errors of the individual coordinates, global and local 3D indices, and their absolute confidence ellipsoid elements determining 3D accuracy of some chosen replaced points; the numbers in front of the slash belong to 1976; the numbers after the forward slash belong to 2014. In 2007 the points No. 2, 3, 30, 38, 104, 105, and 227 were re-stabilized due to small earthworks for the purpose of some preparation work for the future reclaiming of the mining territory of Košice-Bankov. The values of the deformation vectors confirm the fact that the presented theory of the deformation vector specific solution is suitable and variable adapted to various damages of geodetic networks [7]. In geodetic practice, however, there are cases where the values of deformation vectors need not mean any displacement of the geodetic network points (movement around the point of the geodetic network). Although the geodetic network points were adjusted according to the conventional method using the Gauss-Mark model, the deformation vector values may be loaded by the accumulation of some measurement errors. For this reason, when evaluating the

[2] Deformation survey on the monitoring station of Košice-Bankov without the reclaiming work intervention was finished in the autumn 2014.

Point	m_X (mm)	m_Y (mm)	m_Z (mm)
2	15.7/16.4	32.9/45.5	12.5/72.4
3	14.8/34.3	27.2/58.9	30.5/69.1
30	21.1/25.6	26.5/24.1	45.5/32.7
38	16.6/14.9	16.3/8.1	20.1/18.4
104	18.2/41.3	34.1/69.0	55.4/78.0
105	28.2/31.6	17.1/21.1	9.9/17.8
227	20.0/16.9	8.5/4.7	10.9/10.8

Table 1.
Mean errors (1976/2014).

Point	a_i (mm)	b_i (mm)	φ_{a_i} (gon)	f
2	49.9/51.0	5.9/5.1	172.303/172.695	1.8818/0.9
3	40.8/32.5	12.3/3.7	172.704/179.151	0.6985/0.8862
30	43.0/45.9	18.2/21.5	160.340/160.058	0.5767/0.5316
38	23.5/28.1	21.8/22.4	40.966/41.203	0.0723/0.2028
104	47.5/79.2	24.0/9.9	211.146/217.148	0.4947/0.875
105	42.8/42.4	15.3/17.6	370.337/370.624	0.6425/0.5849
227	28.8/25.2	8.1/10.9	19.634/19.781	0.7188/0.5675

Table 2.
Absolute confidence ellipse elements (α = 0.05) (1976/2014).

Rank rk$(\Sigma_{\hat{C}})$	Track tr$(\Sigma_{\hat{C}})$ [mm^2]	Determinant $det(\Sigma_{\hat{C}})$	Average mean error $\sigma_{\hat{C}_{pr}}$ [mm]	Norm $nor(d_{\hat{C}})$ [mm]
14/14	7041.901/ 7041.054	$2.869.10^{25}$/ $2.869.10^{25}$	22.428/22.971	124.218/ 126.155

Table 3.
Global indices (1976/2014).

Point	Mean 3D error σ_p (mm)	Mean coordinate error σ_{XYZ} (mm)
2	36.4/37.5	25.7/18.8
3	30.9/28.4	21.8/24.5
30	33.9/33.0	23.9/23.0
38	23.3/26.7	16.5/13.8
104	38.6/14.1	27.3/54.9
105	32.9/27.4	23.3/19.9
227	21.7/22.5	15.3/17.4

Table 4.
Local indices (1976/2014).

	Point						
$d\hat{C}$ [mm]	2	3	30	38	104	105	227
	2.4	−2.9	−8.0	6.7	−4.0	0.6	9.7

Table 5.
Deformation vector values (1976/2014).

significance of deformation vectors, it is necessary to test them by means of the global and localization test of a congruence (see Chapter 3.3). After a series of the last geodetic measurements in spring 2014, the deformation vectors at the tested points No. 2, 3, 30, 38, 104, 105, and 227 of the geodetic network of the monitoring station have moved from −4 to 9.7 mm (**Table 5**). 3D mean errors ranged from 14.1 to 38.6 mm, and the mean coordinate errors were from 13.8 to 54.9 mm (**Table 4**). In autumn 2014 all points of the monitoring station of Košice-Bankov were destroyed by the reclaiming work, i.e., the reference points were removed and the object points were backfilled by a secondary imported soil.

3.3 Global test of the geodetic network congruence

Significant stability, respectively instability of the geodetic network points, is rejected or not rejected by verifying the null-hypothesis H_0 and also other alternative hypothesis [11, 12, 17, 34].

$$H_0 : d\hat{\mathbf{C}} = 0; \quad H_\alpha : d\hat{\mathbf{C}} \neq 0 \tag{18}$$

where H_0 expresses insignificance of the coordinate differences (deformation vector) between epochs $t_{(0)}$ and $t_{(i)}$.

The test statistics T_G can be used for a global test:

$$T_G = \frac{d\hat{\mathbf{C}}\mathbf{Q}_{d\hat{C}}^{-1}d\hat{\mathbf{C}}^T}{k\bar{s}_0^2} \approx F(f_1, f_2) \tag{19}$$

where $\mathbf{Q}_{d\hat{C}}$ is the cofactor matrix of the final deformation vector $d\hat{\mathbf{C}}$, k is the coordinate number entering the geodetic network adjustment (k=3 for 3D coordinates), and $\bar{s}_0^2$ is the posteriori variation factor common for both epochs $t_{(0)}$ and $t_{(i)}$.

The critical value T_{KRIT} is searched in the tables of F distribution (the Fisher-Snedecor distribution) according to the degrees of freedom $f_1 = f_2 = n - k$ or $f_1 = f_2 = n - k + d$, where n is number of the measured values entering into the network adjustment and d is the network defect at the network free adjustment. Through the use of methods, MINQUE is $s_0^{2_{t(0)}} = s_0^{2_{t(i)}} = \bar{s}_0^2 = 1$ [2, 11, 12, 16, 17, 34]. Test statistics T should be compared to critical test statistics T_{KRIT}.T_{KRIT} is found in the distribution Tables F according to the degrees of freedom of the geodetic network.

When comparing test statistics, there may be two cases:

1. $T_G \leq T_{KRIT}$: The null-hypothesis H_0 is accepted. That is, the differences in coordinate values (i.e., deformation vectors) are not significant.

2. $T_{G>}T_{KRIT}$: The null-hypothesis H_0 is refused. This means that the differences in coordinate values (deformation vectors) are statistically significant. In this

Point	$T_{G(i)}$	$< \leq >$	F	Notice
2	1.297	<	3.724	Deformation vectors are not significant
3	3.7236	≤		
30	3.501	<		
38	3.7237	≤		
104	2.871	<		
105	1.403	<		
227	2.884	<		

Table 6.
Test statistics results of the geodetic network points at the monitoring station of Košice-Bankov.

case, it can be stated that the deformation occurred with the level α of a reliability.

Table 6 presents the global tested results of the geodetic network congruence.

4. Subsidence in GIS for reclaiming mining landscape

The Geographical Information Systems (GIS) of the mining landscape of Košice-Bankov is based on the following key points [25]: basic and simple presentation of geodata, management of the basic database, and wide availability of information. The best feasible solution for the implementation of the GIS project is the Free Open Source software applications, which are easily available on the Internet. The general function of the Free Open Source software application is the viability of free code and data sources via HTTP and FTP protocols located on the project website. Other features of the Free Open Source range include ease of use, access to data and information, centralized system configuration, modular things, and any Open Source platform (depending on PHP, MySQL, and ArcIMS ports) [7, 25, 35–39]. The network application MySQL is currently the most advantageous database system on the Internet and was also applied to the deformation survey outputs from the monitoring station of Košice-Bankov.

Whole database part in GIS for the subsidence of Košice-Bankov in all applications was processed into the MySQL database (**Figure 5**). 3D model of the subsidence of Košice-Bankov with the multilayered GIS applications has been implemented into the reclaiming plan of the mining territory of Košice-Bankov for the needs of the municipality of the city of Košice.

Given the fact that extraction of magnesite mineral has been completed at the mine of Košice-Bankov and this mine is abandoned since the end of the 90 years of the twentieth century and the investigation concluded that the Košice-Bankov with the huge subsidence is stable at the end of the deformation investigation, the municipality of the city of Košice (Department of land planning of the city) has adopted a definitive plan for reclaiming this mining landscape. Numerical and graphical analyses of the results from the long-term geodetic measurements of the deformations in the subsidence at the monitoring station of Košice-Bankov with their subsequent testing analyses of the deformation vectors confirmed the stability not only in the subsidence but also in the surrounding mining landscape. The subsidence and other mining earthworks of huge dimensions destroyed the entire surroundings of the mining plant (mining tailings piles, various excavations in the

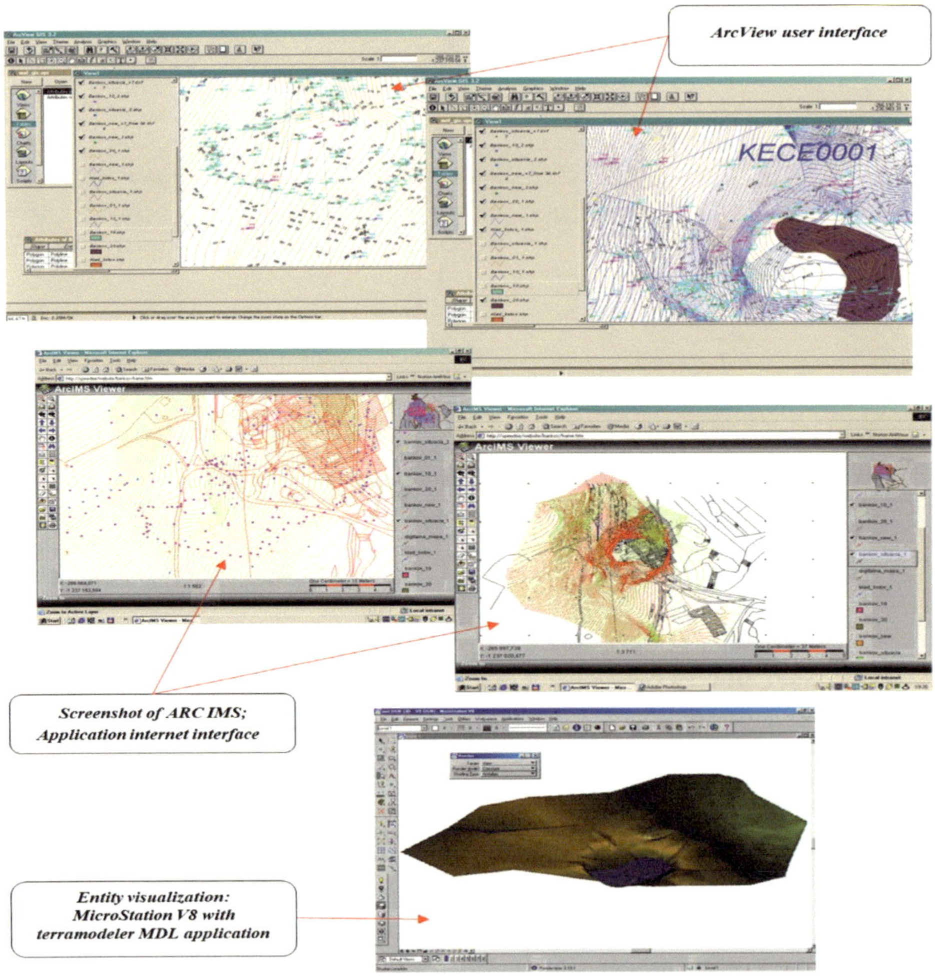

Figure 5.
ArcView user interface visualization of the subsidence.

working of the earth surface, and other mining works in the surroundings of the former mine plant) were gradually filled with imported secondary soil. Based on the results of the extensive geodetic measurements of deformations of the mining subsidence and its surroundings in the destroyed mining area of Košice-Bankov, reclaiming works began at the beginning of this century. Some recent reclaiming works were completed in the summer of 2016.

In the territory of the former large subsidence, the new forest park was built as an environmental forest greenery in the part of the Košice-Bankov urban recreation and touristic area of the city of Košice. The subsidence was filled with imported natural and secondary materials from many construction works and earthworks in the city of Košice and its surroundings. Given that the subsidence was of huge proportions, such sporadic embankment works took too long, more than 10 years. After the embankment and other earthworks were completed, the new forest park was planted in the area of the former subsidence (**Figure 6**). Especially birch was planted. Birch trees are known for their rapid growth, and their root system is not demanding on the underlying soil. At present, the birch grove represents almost 5 years of a healthy forest park. The reconstruction of the recreational and touristic area of Košice-Bankov was completed in spring 2016 (**Figure 7**). The mining tailing piles and the entire ruined area of the former mining plant were also reclaimed.

DOI: http://dx.doi.org/10.5772/intechopen.91461

Figure 6.
The subsidence of Košice-Bankov after reclaiming; panoramic view—Summer 2016. Solar collectors on the places of the former mining tailing piles; new forest park in the background on the places of the former subsidence.

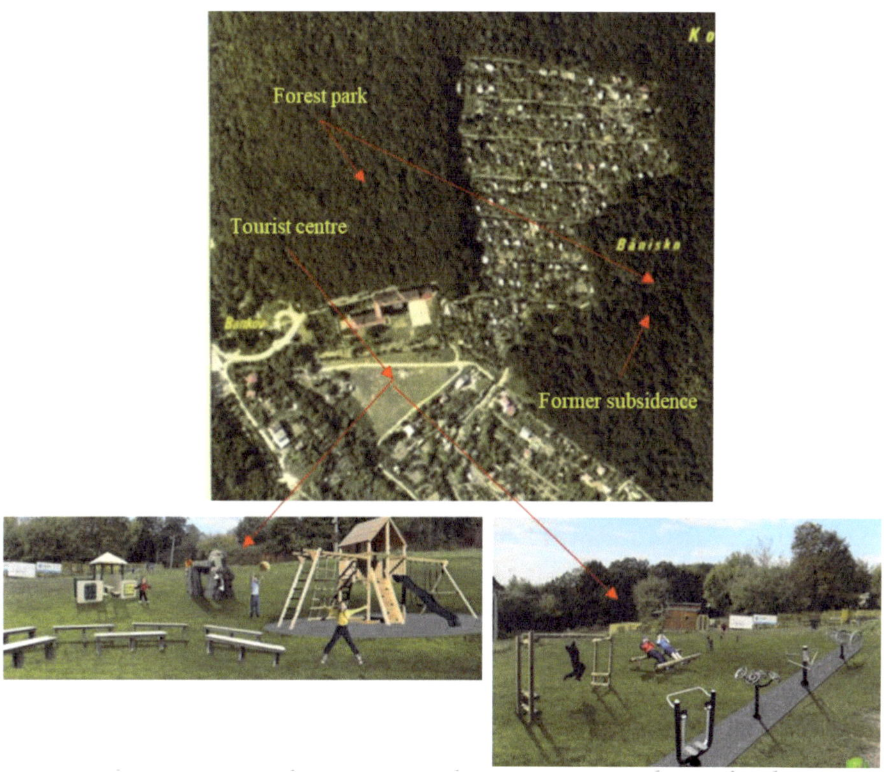

Figure 7.
The reconstructed recreation and touristic zone and revitalized forest park after reclaiming the mining landscape of Košice-Bankov.

Many solar collectors have been built on the places of the former mining tailing piles, which contribute to the renewable sources of electricity for the city of Košice (**Figure 6**).

5. Conclusions

The determination of the deformation vectors from the differences between the adjusted geodetic point coordinate vectors obtained from at least two monitoring epochs is readily achievable if the original geometric and data structure of the geodetic network of the monitoring station between the individual monitoring epochs is strictly preserved. The presented research study provides both the theoretical and practical results of the possibility of solving deformation vectors in the geodetic network of the monitoring station in the case of its geometrical and data structure violation during the period of monitoring the movements of the earth surface, i.e., if the points of the geodetic network were damaged or completely destroyed between the different epochs of measurements, i.e., the geodetic network was nonhomogenous. The deformation vectors solved in accordance with the proposed specific theory of solving monitored deformations of the earth surface in the case of disturbance of the geodetic network homogeneity of the monitoring station provide via 3D models in GIS a reliable idea of spatial changes in the coordinates of the geodetic network points. The proposed theory of the specific deformation vector solution leads to a reliable support in investigation of various deformations of the earth surface, such as mining subsidence, landslides, geotectonic (recent movements), movements of dams, and other important building objects.

The largest values of the deformation vectors given in **Table 5** occurred at points No. 30 and No. 227. However, due to the fact that the deformation vectors tested at these points were not significant according to the test statistics, we can declare these points as static. The mentioned study case from the mining territory of the abandoned magnesite mine of Košice-Bankov confirmed the availability and applicability of the presented specific theory in the solution of the deformation vector in the deformation monitoring in the mining subsidence, where several violations of the geometric and data structure (homogeneity) of the geodetic network of the monitoring station occurred. Despite the validated method for the specific solution of the deformation vector in the geometric and data inhomogeneity of the geodetic network at the monitoring station, it should be stated that it would be preferable to maintain the homogeneity of the geodetic network during the whole monitoring period. Maintaining the homogeneity of the data of the geodetic network structure can ensure the permanent stabilization of the network points and the correct technically and physically implemented protection of the whole monitoring station from unexpected external interventions into such a station.

3D model situations of the mining subsidence in GIS platform from the mining territory of Košice-Bankov were delivered to the municipality of the city of Košice (especially for the Department of Land Planning and Chief Architect of the City) to deal with a spatial planning for the future environmental reclaiming of this abandoned mining region, such as the magnesite mines of Košice-Bankov. The analysis of the deformation vectors at the geodetic network points of the monitoring station located in the mining subsidence and in the surrounding mining territory of the abandoned magnesite mine of Košice-Bankov was important in defining and specifying the subsidence edges and the subsidence zones with a number of dangerous cracks and fissures. The very precise identification of 3D position of such delimitation of the mining subsidence constituted the basic document for the plan of the municipality of the city of Košice for the reclaiming of the entire devastated mining landscape of Košice-Bankov. The revitalization of the Košice-Bankov recreational and tourist zone with the adjacent urban forest park has been achieved through the comprehensive reclaiming that devastated mining landscape. The variability of 3D models of the mining subsidence allowed a wide spectrum at modeling of natural and also industrial disasters in the former mining territory of Košice-Bankov.

3D models of the mining subsidence in GIS are useful tools for many reclaiming works in restoring ecosystems with some essential elements of the security measures against the possible and unforeseen consequences of former mining activities for the protection of the health and life of people moving in various mining areas.

Acknowledgements

This research study was supported by the Slovak Research and Development Agency under the contract No. SK-CN-RD-18-0015 and by the European Union within the Interreg ENI CBC Programme under the contract No. HUSKROUA/1702/8.1/0065 and by the Scientific Grant Agency of the Ministry of Education, Science, Research and Sports of the Slovak Republic under the contract No. VEGA 1/0839/18.

Author details

Vladimír Sedlák
Pavol Jozef Šafárik University in Košice, Slovakia

*Address all correspondence to: vladimir.sedlak@upjs.sk

References

[1] Knothe S, editor. Forecasting the Influence of Mining (in Polish). Katowice: Śląsk Publishing House; 1984

[2] Whittaker BN, Reddish DJ, editors. Subsidence: Occurrence, Prediction and Control. 3rd ed. Amsterdam: Elsevier; 1989. p. 528

[3] Cui X, Miao X, Wang J, Yang S, Liu H, Hu X. Improved prediction of differential subsidence caused by underground mining. International Journal of Rock Mechanics and Mining Sciences. 2000;**37**(4):615-627. DOI: 10.1016/S1365-1609(99)00125-2

[4] Díaz-Fernández ME, Álvarez-Fernández MI, Álvarez-Vigil AE. Computation of influence functions for automatic mining subsidence prediction. Computational Geosciences. 2010;**14**(1):83-103. DOI: 10.1007/s10596-009-9134-1

[5] Djamaluddin I, Mitani Z, Esaki T. Evaluation of ground movement and damage to structures from Chinese coal mining using a new GIS coupling model. International Journal of Rock Mechanics and Mining Sciences. 2011;**48**(3):380-393. DOI: 10.1016/j.ijrmms.2011.01.004

[6] Kratzsch H, editor. Mining Subsidence Engineering. Berlin: Springer-Verlag; 1983

[7] Sedlák V. Mathematical testing the edges of subsidence in undermined areas. Journal of Mining Science. 2014;**50**(3):465-474. DOI: 10.1134/S1062739114030089

[8] Bauer RA. Mine subsidence in Illinois: Facts for homeowners. In: Circular 569 [Internet]. Champaign: Illinois State Geological Survey; 2006. p. 20. Available from: https://www2.illinois.gov/iema/Mitigation/Documents/Link_Mine_Subsidence_Facts_Homeowners.pdf [Accessed: 24 September 2016]

[9] Colorado Geological Survey: Mine Subsidence [Internet]. 2016. Available from: http://coloradogeologicalsurvey.org/geologic-hazards/subsidence-mine/ [Accessed: 26 September 2016]

[10] Donnelly LJ, Reddish DJ. The development of surface steps during mining subsidence: "Not due to fault reactivation". Engineering Geology. 1994;**36**(3–4):243-255. DOI: 10.1016/0013-7952(94)90006-X

[11] Sedlák V, editor. Modelling Subsidence Development at the Mining Damages. Košice: Štroffek; 1997. p. 52

[12] Sedlák V. Modelling subsidence deformations at the Slovak coalfields. Kuwait Journal of Science & Engineering. 1997;**24**(2):339-349

[13] Alehossein H. Back of envelope mining subsidence estimation. Australian Geomechanics. 2009;**44**(1):29-32

[14] Can E, Mekik Ç, Kuşçu Ş, Akçın H. Monitoring deformations on engineering structures in Kozlu Hard Coal Basin. Natural Hazards. 2013;**65**(3):2311-2330. DOI: 10.1007/s11069-012-0477-x

[15] Jung HC, Kim SW, Jung HS, Min KD, Won JS. Satellite observation of coal mining subsidence by persistent scatterer analysis. Engineering Geology. 2007;**92**(1):1-13. DOI: 10.1016/j.enggeo.2007.02.007

[16] Sedlák V, Kunák L, Havlice K, Šadera M. Modelling deformations in land subsidence development at the Slovak coalfields. Survey Ireland. 1995;**12**(13):25-29

[17] Sedlák V. Measurement and prediction of land subsidence above

longwall coal mines, Slovakia. In: Borchers WJ, editor. Land Subsidence: Case Studies and Current Research. Belmont: U.S. Geological Survey; 1998. pp. 257-263

[18] Cai J, Wang J, Wu J, Hu C, Grafarend E, Chen J. Horizontal deformation rate analysis based on multiepoch GPS measurements in Shanghai. Journal of Surveying Engineering. 2008;**134**(4):132-137. DOI: 10.1061/(ASCE)0733-9453(2008)134:4(132)

[19] Can E, Mekik Ç, Kuşçu Ş, Akçın H. Computation of subsidence parameters resulting from layer movements post-operations of underground mining. Journal of Structural Geology. 2013;**47**: 16-24. DOI: 10.1016/ j.jsg.2012.11.005

[20] Hu LY. Gradual deformation and iterative calibration of Gaussian-related stochastic models. Mathematical Geology. 2000;**32**(1):87-108. DOI: 10.1023/A:1007506918588

[21] Lü WC, Cheng SG, Yang HS, Liu DP. Application of GPS technology to build a mine-subsidence observation station. Journal of China University of Mining and Technology. 2008;**18**(3):377-380. DOI: 10.1016/S1006-1266(08)60079-6

[22] Marschalko M, Fuka M, Treslin L. Measurements by the method of precise inclinometry on locality affected by mining activity. Archives of Mining Sciences. 2008;**53**(3):397-414

[23] Ng AH, Ge L, Zhang K, Chang HC, Li X, Rizos C, et al. Deformation mapping in three dimensions for underground mining using InSAR - Southern highland coalfield in New South Wales, Australia. International Journal of Remote Sensing. 2011;**32**(22): 7227-7256. DOI: 10.1080/ 01431161.2010.519741

[24] Sedlák V. GPS measurement of geo-tectonic recent movements in east Slovakia. In: Proceedings of the 6th International Symposium on Land Subsidence–SISOLS 2000; 24–29 September; Ravenna. Ravenna: C.N.R.; 2005. Vol. II, 2000. pp. 139-150

[25] Sedlák V. Possibilities at modelling surface movements in GIS in the Košice depression, Slovakia. RMZ-Materials and Geoenvironment. 2004;**51**(4): 2127-2133

[26] Wright P, Stow R. Detecting mining subsidence from space. International Journal of Remote Sensing. 1999;**20**(6): 1183-1188. DOI: 10.1080/ 014311699212939

[27] Koníček P, Souček K, Staš L, Singh R. Long-hole destress blasting for rockburst control during deep underground coal mining. International Journal of Rock Mechanics and Mining Sciences. 2013;**61**:141-153. DOI: 10.1016/j.ijrmms.2013.02.001

[28] Strazalowski P, Scigala R. The example of linear discontinuous deformations caused by underground extraction. Transection of the VŠB - Technical University of Ostrava, Civil Engineering Series. 2005;**V**(2):193-198

[29] Li PX, Tan ZX, Deng KZ. Calculation of maximum ground movement and deformation caused by mining. Transactions of the Nonferrous Metals Society of China. 2011;**21**(3): s562-s569. DOI: 10.1016/S1003-6326(12)61641-0

[30] Christensen R. General Gauss-Markov models. In: Christensen R, editor. Plane Answers to Complex Questions/the Theory of Linear Models. 4th ed. New York: Springer; 2011. pp. 237-266

[31] Gene H, Golub GH, Van Loan ChF, editors. Matrix Computations. 4th ed. Baltimore: JHU Press; 2013. p. 756

[32] Groß J. The general Gauss-Markov model with possibly singular dispersion

matrix. Statistical Papers. 2004;**45**(3): 311-336. DOI: 10.1007/BF02777575

[33] Lindgren F, Ruel H, Lindström J. An explicit link between Gaussian fields and Gaussian Markov random fields: The stochastic partial differential equation approach. Journal of the Royal Statistical Society, Series B: Statistical Methodology. 2011;**73**(4):423-498. DOI: 10.1111/j.1467-9868.2011.00777.x

[34] Lehmann EL, Romano JP, editors. Testing Statistical Hypotheses. 3rd ed. New York: Springer; 2008. p. 786

[35] Blachowski J. Application of GIS spatial regression methods in assessment of land subsidence in complicated mining conditions: Case study of the Walbrzych coal mine (SW Poland). Natural Hazards. 2016;**84**(2):997-1014. DOI: 10.1007/s11069-016-2470-2

[36] Gallay M, Kaňuk J, Hochmuth Z, Meneely JD, Hofierka J, Sedlák V. Large-scale and high-resolution 3-D cave mapping by terrestrial laser scanning: A case study of the Domica Cave, Slovakia. International Journal of Speleology. 2015;**44**(3):277-291. DOI: 10.5038/1827-806X.44.3.6

[37] Kaňuk J, Gallay M, Hofierka J. Generating time series of virtual 3-D city models using a retrospective approach. Landscape and Urban Planning. 2015;**139**:40-53. DOI: 10.1016/j.landurbplan.2015.02.015

[38] Yang KM, Xiao JB, Duan MT, Pang B, Wang YB, Wang R. Geo-deformation information extraction and GIS analysis on important buildings by underground mining subsidence. In: Proceedings of the International Conference on Information Engineering and Computer Science–ICIECS 2009. 19–20 December, 2009; Wuhan. Wuhan: IEEE; 2009. p. 4

[39] Yang KM, Ma JT, Pang B, Wang YB, Wang R, Duan MT. 3D visual technology of geo-deformation disasters induced by mining subsidence based on ArcGIS engine. Key Engineering Materials. 2012; **500**:428-436. DOI: 10.4028/www.scientific.net/KEM.500.428

Chapter 9

Coal Burst: A State of the Art on Mechanism and Prevention from Energy Aspect

Xiaohan Yang

Abstract

Coal burst continues to be one of the most catastrophic safety hazards faced by future mining as the stress environment will be more complicated with the increase of mining depth. Many chief coal mining countries including Poland, Czech Republic, the U.S., China, and Australia have experienced fatal accidents caused by coal burst and conducted comprehensive research on the driving forces and solving technologies related to coal burst. In this chapter, the research outcomes of the mechanism, risk evaluation, risk monitoring, and prevention of coal burst are reviewed, which is helpful for mining researchers and engineers to understand and control the safety hazards caused by coal burst, and, hence, to achieve sustainable and safe mining.

Keywords: coal burst, underground mining, mining safety, dynamic hazards, rock mechanics

1. Introduction

Coal burst, which refers to the violent and catastrophic failure of coal, is a serious safety hazard for underground coal mines, and it has attracted intensive research interests from mining and geological scholars [1]. In 1738, the first recorded coal burst took place in England [2, 3]. Since then, both the frequency and severity of coal burst increased with mining depth [2, 4, 5]. As shown in **Table 1**, coal burst has been a serious security issue that many countries face for decades. Coal burst has been recognized as a serious risk for Australia's underground coal mines following a fatal coal burst accident at the Austar Coal Mine [6, 7]. Because of lacking coal burst experience, it is difficult to find mature theories and technologies in Australian to explain, predict, monitor, or control coal burst. It is an urgent task to develop a coal burst risk assessment methodology and prevention technology for Australian coal mines. Extensive study has been conducted around the mechanism, prediction, and prevention of coal burst [5] by scholars around the world. Some necessary conditions of coal burst such as stiffness, dynamic load, and mechanical property are found based on previous decades' research.

In terms of energy, coal burst is the energy accumulation and releasing process of a coal body. Coal burst monitoring, such as acoustic emission, electromagnetic radiation, micro-seismic, infrared, and other methods, is the monitoring of different energy forms released during coal burst [8, 9]. The cause of the coal ejection and roadway destruction is the elastic energy stored in the coal [10]. Therefore,

Country/region	Time period	Number of coal bursts	Number of fatalities	Reference
Czech Republic/ Poland	1983–2003	190	122	
Ruhr, Germany	1973–1992	50	27	[4]
USA	1943–2003	–	78	
USA	1983–2013	337	20	[21]
Mainland China	1933–1996	4000	400	[5]
Mainland China	2006–2013	>35	>300	[36]

Table 1.
Coal burst occurrence and fatalities by country/region [7].

it is significant to have an understanding of energy release mode in the coal burst process, especially the magnitude of coal burst energy. Coal burst is regarded as a dynamic disaster since it is shown in many studies that coal burst is closely related to dynamic load [11]. It is believed that hard rock is more prone to violent failure than soft rock [12]. Due to the difference in physical and mechanical properties, different coal seams have a different coal burst propensity. Therefore, changing coal mechanical property is a promising method for coal burst mitigation. Water infusion can mitigate coal burst propensity through increasing moisture content of coal [13]. In this chapter, the coal burst driving forces, solving techniques, and monitoring methods are reviewed from energy aspects.

2. Potential driving factors

2.1 Mining depth

Mining depth has been identified as an important factor for the formation of coal burst. According to the analysis of coal burst cases in Poland and China, LM Dou found that the first coal burst accident in coal mines generally happened when mining depth approached 350 m and the frequency and severity of coal burst sharply increases with mining depth changing from 350 to 600 m [14]. Iannacchione and Zelanko found that nearly all coal bursts in the main coal fields of the U.S. occurred at depths greater than 300 meters, and most were at depths exceeding 400 m [15]. The contribution of mining depth to coal burst mainly results from the increasing gravitational stress. More strain energy will be stored in the coal under high gravitational stress condition [16]. Besides, for coal mines in China and the U.S., hard sandstone roof seems the common geological feature for deep mining, which can further result in a large accumulation of energy or a catastrophic dynamic load [17, 18]. The potential influence of hard roof (roof stiffness) also will be discussed in another section of this paper. The mining depths of two coal mines with coal burst accident in Australia are both around 500 m [19]. Hence, the strain energy accumulation led by high gravitational stress plays an important role in the formation of coal burst accidents that happened in Australia as the mining depth of these coal mines is already beyond the mining depth of majority of burst accidents revealed by international research.

More seriously, almost all coal mines in Australia have plans for deeper mining, which means the stress environments will be more complicated and more energy will be stored in coal seams [20].

2.2 Geological structures

It has been shown by numerous studies that the complicated geological structures caused by folds, faults, and coal seam thickness variation have a noticeable influence on the coal burst occurrence [21]. Dou et al. found that 72% coal burst accidents in Longfeng Colliery were related to faults [16]. The numerical study conducted by Chen et al. found that stress will concentrate near the coal face when the coal face approaches fault [22]. Mark found that coal burst accidents in the U.S. have a close relationship with faults [23]. Folds, which are created by compressional tectonic stress, may have high residual tectonic stress in the geological structures. Through the stress regression analysis of Huanghuiyan Colliery, Jiang et al. found that stress concentration tends to exit at the area near syncline axis [24]. The influence of geological structures on stress distribution is shown as **Figure 1**.

2.3 Surrounding rocks' stiffness

Stiffness of the surrounding rocks is one of the main factors giving rise to coal burst. Bieniawski found that rock samples are more prone to violent failure under the loading machine with high stiffness. The uniaxial compression tests of sample composed by coal and rock found that most elastic energy is stored in the coal part of the compound sample and the burst potential of the sample is positively related to the thickness of the rock part [25, 26]. Through theoretical analysis, Yang found that energy will flow from high stiffness material to low stiffness material [17]. Hence, the high stiffness of surrounding rocks will enhance the energy accumulation in coal seam. In addition, as shown in **Figure 2**, the strength of coal tends to have rapidly decreased under the high stiffness environment [27]. Generally, the high stiffness environment is related to the heavy and hard sandstone layer above the coal seam [28]. Sometimes, the thickness of sandstone layer can reach tens or even hundreds of meters [16].

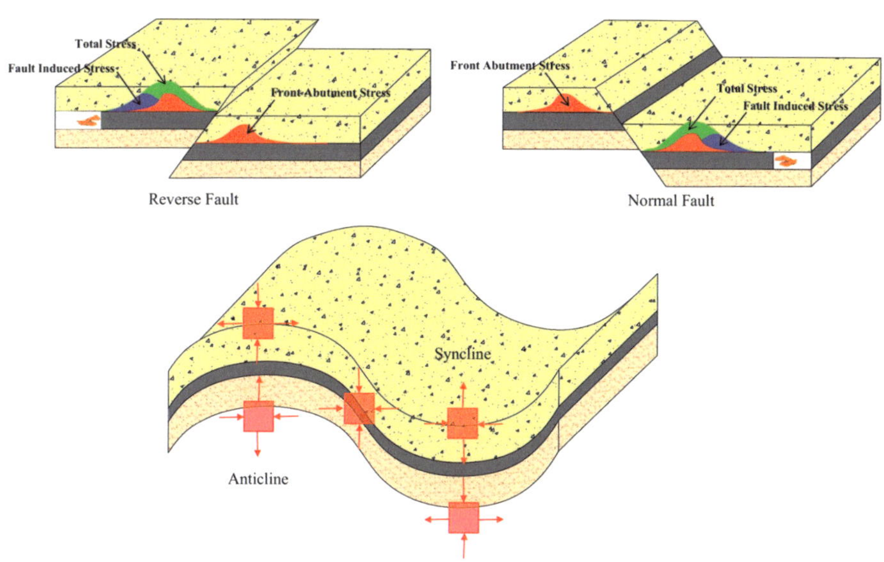

Figure 1.
Stress concentration caused by geological structures.

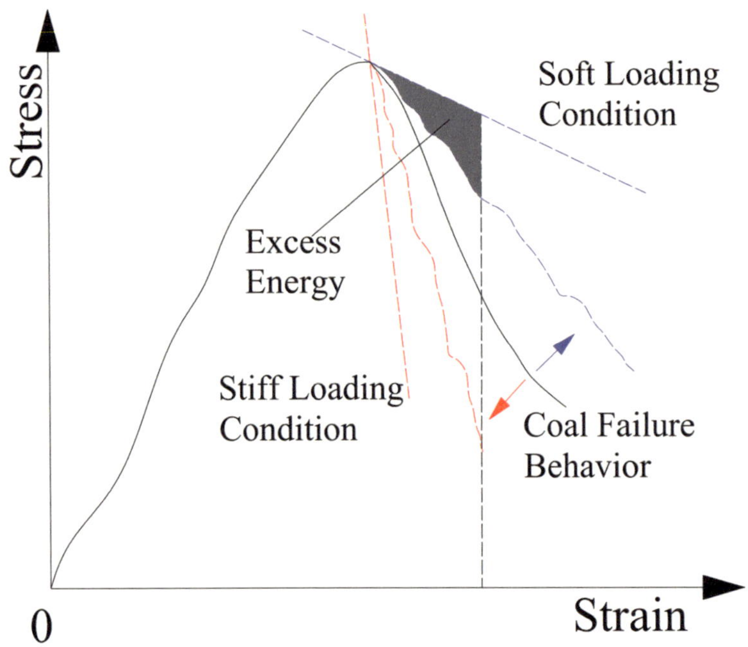

Figure 2.
Effect of stiffness of the loading system on the behavior of coal failure [17].

2.4 Micro-seismicity

Micro-seismicity refers to the regional small-scale seismic events that are undetectable by earthquake monitoring stations due to their small-scale energy compared with earthquakes. However, for underground coal mines, the energy released by micro-seismicity also is an important energy source for coal burst formation. Intensive micro-seismicity has been observed in most coal mines with high bursts risk in Poland, China, and the U.S. [29–31]. Micro-seismicity can be detected and located by specific micro-seismic monitoring apparatus. Deep research has been made by many researchers on the monitoring of dynamic load and identifying high burst potential areas through micro-seismic monitoring [32–34].

3. Previous mechanism

The study of the coal burst mechanism aims to explain the causes of coal burst from two perspectives: force source and coal's physical properties. As a type of coal failure, coal burst should meet the conditions of coal failure. That is, the stress loaded on coal exceeds the strength of coal when coal burst occurs, which is named strength theory by some scholars [16]. Satisfying strength theory is one of the conditions required by coal burst. Under static loading condition, coal burst does not always happen when the ultimate strength is reached. It has been pointed out that coal strength will change under dynamic load. Research has shown that the coal failure behavior is affected by loading rate as well [35]. In the actual situation, the strength theory of coal burst becomes more complex as the coal body is under the collective effect of static load (overburden weight) and dynamic load. Dou et al. [14] studied the dynamic load required by coal burst at different static load levels. Through a series of follow-up studies, LM Dou put forward the dynamic and static load superposition theory of coal burst [36, 37]. The strength theory of coal burst

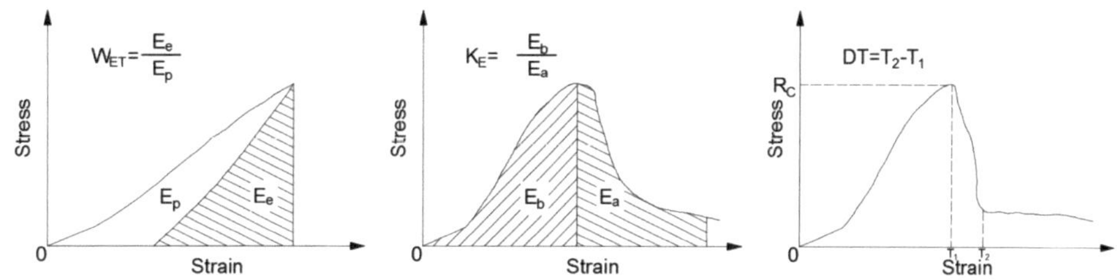

Figure 3.
Schematic diagram of coal burst propensity index [40]. (a) Determination of W_{ET} and (b) determination of K_E.

under dynamic load should be based on the dynamic strength of coal. Cook found that marble only has violent failure when the stiffness of the test machine is greater than the stiffness of the specimen [2, 16]. The compressive experiment of samples composed of coal and rock showed that violent failure always occurred in the layer with minimum stiffness [25, 38]. That is, the necessary condition for coal burst of a pillar or rib is that the stiffness of the roof and floor is greater than that of the coal seam. In most cases, the stiffness of coal seam is minimal relative to roof and floor. That is, coal failure in coal mines generally meet stiffness conditions.

It is found that the post-failure curve of hard rock is steeper than that of soft rock. This means that hard rock is more likely to fail instantaneously. Bieniawski et al. [39] believe that hard rock is much more prone to violent rupture than soft rock. It is necessary to explain that the hard rock and soft rock here are classified in terms of strength. Bieniawski proposed two indices, elastic strain energy index (W_{ET}) and bursting energy index (K_E), to measure the rock burst tendency of different rocks. As shown in **Figure 1**, elastic strain energy index is the ratio between elastic energy (E_e) and plastic energy (E_p) when the specimen is loaded to at least 80% of the strength and then unloaded [2]. K_E is the ratio between E_b and E_a [2]. E_b represents the energy storage before strength while E_a means deformation energy consumed after the peak value. It is proved by in suit and experimental data that coal with high W_{ET} and K_E value has a high tendency for violent failure [2, 4, 25]. These two indices describe the proportion of elastic energy during coal burst. Different rock types have different burst tendency and different energy storage and releasing behavior. Due to the difference in physical and mechanical properties, the W_{ET} and K_E values of different coal seams vary widely as well. Theoretically speaking, coal has no burst ability when the W_{ET} and K_E values are low enough. The ability or property of coal burst is called coal burst propensity by Chinese scholars. Four indices including W_{ET} and K_E are summarized as coal burst propensity indices by Chinese scholars and have become a good indicator of coal burst risk of different coal seams. Coal burst propensity index describes the proportions of different energies. The successful application of the coal burst propensity index method indicates that elastic energy and coal burst are closely related. Coal has the ability to store and instantly release elastic energy in the premise of coal burst (**Figure 3**).

4. Prevention methods

4.1 Evaluation

Based on the analysis of stress-strain curve of coal specimens under uniaxial compression stress, several special indices are published by different researchers to evaluate coal burst propensity. Russian and Poland coal mines adopt elastic strain energy index and bursting energy index to evaluate coal burst propensity [2, 4]. Zhang et al.

believe that the duration of failure process is the comprehensive reflection of energy accumulation and dissipation characteristics of coal [41]. They propose a dynamic failure time to evaluate coal burst propensity. Based on the correlation analysis of mass data, Qi et al. conclude that uniaxial compression strength of coal is a proper index of coal burst propensity evaluation as well [42]. In 2010, the China Coal Industry Association summarized these four indices as bursting liability indices of coal and published the standard test method of these four indices. Some researchers adopt these four indices to evaluate the burst propensity of rocks as well. It is has been proved by Russian, Poland, and Chinese experience that these four indices are good indicators to define the burst risk of coal seam. Besides, LM Dou et al. combined geological conditions and technical settings of mining together and proposed comprehensive index method based on the coal burst research in China [16].

4.2 Monitoring

Minimizing the safety risk caused by failure of instability rock/coal is an urgent and essential task for underground mines. Similar with the instantaneous failure of other brittle materials such as rock, concrete, and metal, the coal burst process is always associated with the release of rich geophysical signals including acoustic emission (AE) [43], micro-seismic [32] and electromagnetic radiation [44]. It is demonstrated by decades of research and in-field application that micro-seismic monitoring technology has a promising ability to locate potentially violent rock failure. Micro-seismic monitoring is a passive observation of very small-scale earthquakes that occur in the underground as a result of human activities such as mining, hydraulic fracturing, and underground gas storage. The phenomenon that stressed rock can release micro-level signal was discovered by two researchers of U.S. Bureau of Mines, Obert and Duvall, in 1938 [32, 34]. In the early 1960s, South African researchers developed a 16-channel micro-seismic system with positioning function for rock burst monitoring in gold mines [34]. In 1970, under the sponsorship of the U.S. Bureau of Mines, the Pennsylvania State Rock Mechanics Laboratory conducted a research project to investigate the application of micro-seismic techniques to coal mine safety [45]. Through decades' study of underground micro-seismic for mining operation, micro-seismic system has been a basic and valuable monitoring tool for metal and coal mines worldwide. It provides a continuous and real-time 4D (three dimension location and time) record of seismicity associated with rock failure in the monitoring region.

4.3 Controlling

The widely used coal burst controlling methods include provocative blasting, long-term water infusion, hydro-fracturing, de-stress drilling, and protective seam mining [46]. Dou et al. proposed the intensity weakening theory to guide the coal burst control from the aspect of energy [16]. Based on the energy aspects, the key to coal burst prevention are: (1) softening coal by changing the physical and mechanical properties of coal. The burst tendency or burst scale of soft coal will be mitigated as the energy storage ability of coal has been reduced. The main methods of coal body softening are blasting and water infusion. (2) Transferring stress to deep regions and reducing the stress level of coal, which can reduce energy storage as well. The main methods are pressure relief blasting, roof pre-splitting blasting, roof cutting blasting, protection seam mining, hydraulic roof fracturing, and large diameter pressure relief drilling. (3) Releasing energy by artificially induced coal burst under low stress level. The main methods are pressure relief blasting and large diameter pressure relief drilling.

Conflict of interest

The authors declare no conflict of interest.

Author details

Xiaohan Yang
School of Civil, Mining and Environmental Engineering, University of Wollongong, NSW, Australia

*Address all correspondence to: xy987@uowmail.edu

References

[1] Yang XH, Ren T, He XQ, Tan LH. A review of energy sources of coal burst in Australian coal mines. In: Naj A, Bob K, editors. 2019 Coal Operators Conference, University of Wollongong, Wollongong; 2019

[2] Pan YS. Study on Rockburst Initiation and Failure Propagation. Doctoral Dissertation; Tsinghua University; 1999

[3] Wang WX, Pan CL, Feng T. Fountain rockburst and inductive rockburst. Journal of Central South University. 2000;**7**:129-132

[4] Braeuner G. Rockbursts in Coal Mines and their Prevention. Boca Raton: CRC Press; 1994

[5] Zhou XJ, Xian XF. Research advance on rockburst theory and its engineering application in collieries. Journal of Chongqing University (Nature Science Edition). 1998;**21**:126-132

[6] Bruce H, Jim G. A review of the geomechanics aspects of a double fatality coal burst at Austar colliery in NSW, Australia in April 2014. International Journal of Mining Science and Technology. 2017;**27**:3-7

[7] Justine C, Jan N. Coalburst causes and mechanisms. In: Coal Operators Conference. 2016

[8] Obert L, Duvall W. Use of subaudible noises for the prediction of rock bursts. In: Technical Report Archive & Image Library. 1942

[9] Zakupin A, Bogomolov L, Mubassarova V, Kachesova G, Borovsky B. Acoustic emission and electromagnetic effects in loaded rocks. In: Sikorski W, editor. Acoustic Emission. Croatia: IntechOpen; 2012

[10] Tan YA, Sun GZ, Guo Z. A composite K_{rb} criterion for the ejection characteristics of the burst rock. Scientia Geologica Sinica. 1991:193-200

[11] Dou LM, He J, Cao AY, Cai W, Li ZL, Zhang JW. Mechanism and prevention methods discussion on coal mine rock burst induced by dynamic load. In: High Level Academic Forum of Fifty Anniversary of China National Coal Association, Beijing. 2012

[12] Bieniawski ZT, Denkhaus HG, Vogler UW. Failure of fractured rock. International Journal of Rock Mechanics & Mining Sciences & Geomechanics Abstracts. 1969;**6**:323-330

[13] Frid V. Electromagnetic radiation method water-infusion control in rockburst-prone strata. Journal of Applied Geophysics. 2000;**43**:5-13

[14] Dou LM, Zhao CG, Yang SG, Wu XR. Prevention and Control of Rock Burst in Coal Mine. Xuzhou: China University of Mining and Technology Press; 2006

[15] Mark C. Coal bursts in the deep longwall mines of the United States. International Journal of Coal Science & Technology. 2016;**3**:1-9

[16] Dou LM, He XQ. Theory and Technology of Rock Burst Prevention. 1st ed. Xuzhou: China University of Mining and Technology Press; 2001

[17] Yang XH, Ren T, Alex R, He XQ, Tan LH. Analysis of energy accumulation and dissipation of coal bursts. Energies. 2018;**11**:1816-1827

[18] Agapito JFT, Goodrich RR. Five stress factors conducive to bursts in Utah, USA, Coal Mines. In: 9th ISRM Congress, International Society for Rock Mechanics. 1999

[19] Mine Safety. IIR16-05 Austar Coal Burst. Australia: NSW Department of Industry; 2016

[20] Zhang CG, Ismet C, Bruce H, Ward CR. Assessing coal burst phenomena in mining and insights into directions for future research. International Journal of Coal Geology. 2017;**179**:28-44

[21] Iannacchione AT, Tadolini SC. Occurrence, predication, and control of coal burst events in The US. International Journal of Mining Science and Technology. 2016;**26**:39-46

[22] Chen XH, Li WQ, Yan XY. Analysis on rock burst danger when fully-mechanized caving coal face passed fault with deep mining. Safety Science. 2012;**50**:645-648

[23] Mark C. Coal bursts that occur during development: A rock mechanics enigma. International Journal of Mining Science and Technology. 2017;**28**(1):35-42

[24] Jiang YD, Song HH, Ma ZQ, Ma BJ, Gao LT. Optimization research on the width of narrow coal pillar along goaf tunnel in tectonic stress zone. Journal of China Coal Society. 2018;**43**:319-326

[25] Dou LM, Lu CP, Zong M, Long XTZ, Li ZH. Rock burst tendency of coal-rock combinations sample. Journal of Mining and Safety Engineering. 2006;**23**:43-46

[26] Huang BX, Liu JW. The effect of loading rate on the behavior of samples composed of coal and rock. International Journal of Rock Mechanics and Mining Sciences. 2013;**61**:23-30

[27] Vardar O, Tahmasebinia F, Zhang C, Canbulat I, Saydam S. A review of uncontrolled pillar failures. Procedia Engineering. 2017;**191**:631-637

[28] Whyatt J. Dynamic failure in deep coal: Recent trends and a path forward. In: Singh SP, editor. The 27th International Conference on Ground Control in Mining, Morgantown. 2008

[29] Leśniak A, Isakow Z. Space–time clustering of seismic events and hazard assessment in the Zabrze-Bielszowice coal mine, Poland. International Journal of Rock Mechanics and Mining Sciences. 2009;**46**:918-928

[30] Li ZL, He XQ, Dou LM, Wang GF. Rockburst occurrences and microseismicity in a longwall panel experiencing frequent rockbursts. Geosciences Journal. 2018:1-17

[31] Hallo M. Microseismic surface monitoring network design-sensitivity and accuracy. In: 74th EAGE Conference and Exhibition Incorporating EUROPEC 2012, Copenhagen. 2012

[32] Ge MC. Efficient mine microseismic monitoring. International Journal of Coal Geology. 2005;**64**:44-56

[33] Amitrano D, Arattano M, Chiarle M, Mortara G, Occhiena C, Pirulli M, et al. Microseismic activity analysis for the study of the rupture mechanisms in unstable rock masses. Natural Hazards and Earth System Sciences. 2010;**10**:831-841

[34] Ge MC. Microseismic monitoring in mines. In: Bruce J, editor. Extracting the Science: A Century of Mining Research, Society for Mining, Metallurgy, and Exploration. Collorado, USA: Littleton; 2010

[35] Li HT, Jiang CX, Jiang YD, Wang HW, Liu HB. Mechanism behavior and mechanism analysis of coal sample based on loading rate effect. Journal of China University of Mining and Tecgnology. 2015;**44**:430-436

[36] Jiang YD, Pan YS, Jiang FX, Dou LM, Ju Y. State of the art review on mechanism and prevention of coal bumps in China. Journal of China Coal Society. 2014;**39**:205-213

[37] Dou LM, He J, Cao AY, Gong SY, Cai W. Rock burst prevention methods based on theory of dynamic and

static combined load induced in coal mine. Journal of China Coal Society. 2015;**40**:1469-1476

[38] Mou ZL, Wang H, Peng P, Liu ZJ, Chen YX. Experimental research on failure characteristics and bursting liability of rock-coal-rock sample. Journal of Mining and Safety Engineering. 2013;**30**:841-847

[39] Bieniawski ZT, Denkhaus HG, Vogler UW. Failure of fractured rock. International Journal of Rock Mechanics and Mining Sciences & Geomechanics Abstracts. 1969;**6**:323-341

[40] Yang XH, Ren T, Tan LH, Remennikov A, He XQ. Developing coal burst propensity index method for Australian coal mines. International Journal of Mining Science and Technology. 2018;**28**(5):783-790

[41] Zhang WB, Wang SK, Wu YK, Qu XC. To determine proneness of coal burst by dynamic failure time. Coal Science and Technology. 1986;**14**:31-34

[42] Qi QX, Peng YW, Li HY, Li JQ, Wang YG, Li CR. Study of bursting liability of coal and rock. Chinese Journal of Rock Mechanics and Engineering. 2011;**30**:2736-2742

[43] Kong XG, Wang EY, Hu SB, Shen R, Li XL, Zhan TQ. Fractal characteristics and acoustic emission of coal containing methane in triaxial compression failure. Journal of Applied Geophysics. 2016;**124**:139-147

[44] Song DZ, Wang EY, Liu J. Relationship between EMR and dissipated energy of coal rock mass during cyclic loading process. Safety Science. 2012;**50**:751-760

[45] Hardy HR, Mowrey GL. Study of microseismic activity associated with a longwall coal mining operation using a near-surface array. Engineering Geology. 1976;**10**:263-281

[46] Calleja J, Porter I. Coalburst control methods. In: Coal Operators' Conference, Wollongong. 2016